Cost-effective Titanium Component Technology for Leading-edge Performance

Organizing Committee

M Ward-Close
DERA, UK

C Waters
DERA, UK

S Bishop
MoD, UK

Based on papers presented at a one-day seminar *Cost-effective Titanium Component Technology for Leading-edge Performance* held at The Royal College of Pathologists, London, UK, on 18 November 1999.

IMechE
Seminar Publication

Cost-effective Titanium Component Technology for Leading-edge Performance

Edited by
Professor M Ward-Close

Organized by
The Aerospace Industries Division of
The Institution of Mechanical Engineers (IMechE)

Co-sponsored by
The Titanium Information Group
and
The Propulsion TAC

IMechE Seminar Publication 2000–19

Published by Professional Engineering Publishing Limited for the Institution of Mechanical Engineers, Bury St Edmunds and London, UK.

First Published 2000

ISSN 1357–9193
ISBN 1 86058 316 4

A CIP catalogue record for this book is available from the British Library.

Printed and bound in Great Britain by Antony Rowe Limited, Chippenham, Wiltshire, UK.

Printed by The Cromwell Press, Trowbridge, Wiltshire, UK.

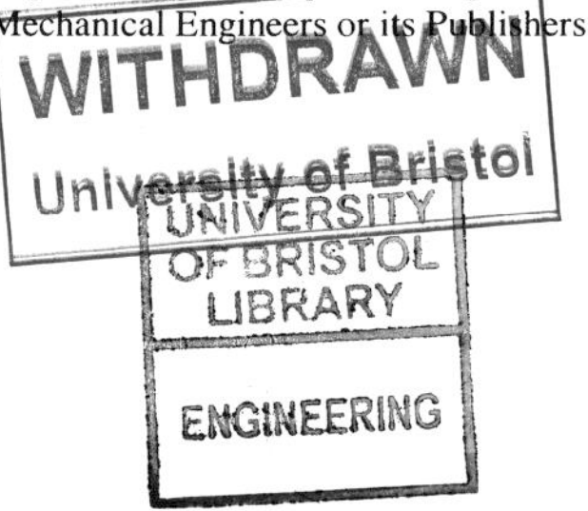

Contents

Related Titles of Interest

Title	Editor/Author	ISBN
Design for Excellence. Engineering Design Conference 2000	Edited by S Sivaloganathan and P T J Andrews	1 86058 259 1
IMechE Engineers' Data Book – Second Edition	C Matthews	1 86058 248 6
Aluminium Materials Technology for Automobile Construction	F Ostermann	0 85298 880 X
Using Powder Metallurgy in Design – Wear, Corrosion, and Fatigue Resistance	IMechE Seminar	1 86058 303 2

For the full range of titles published by Professional Engineering Publishing contact:

Sales Department
Professional Engineering Publishing Limited
Northgate Avenue
Bury St Edmunds
Suffolk
IP32 6BW
UK

Tel: +44 (0)1284 724384
Fax: +44 (0)1284 718692

Foreword

These papers are taken from a one day seminar devoted to titanium applications held at the Royal College of Pathologists, London on 18th November 1999, organized by The Institution of Mechanical Engineers (IMechE) and in collaboration with The Titanium Information Group[1]. The aim of the seminar was to bring together current and new thinking in titanium component design and manufacturing – to demonstrate how titanium technology can offer cost-effective, leading-edge, performance.

Titanium is now price competitive with stainless steel in many applications, and its use is expanding into areas such as sporting goods, automotive, medical, architectural, oil and gas, and land-based military equipment. New processing technologies, such as superplastic forming, have demonstrated that titanium can be cost-effective in applications previously not considered and the amount of information available on both new and traditional manufacturing methods for titanium is increasing rapidly. Techniques such as powder metallurgy, casting, welding, diffusion bonding, extrusion, forging, and machining are all common place now for titanium, and much recent work has concentrated on cost reduction.

The diverse range of papers presented in this volume reflects the growing level of interest in the use of titanium. It is through events such as this symposium that knowledge and understanding of titanium can be spread, enabling titanium to fulfil its promise of joining steel and aluminium as the third major engineering metal for the 21st century.

Malcolm Ward-Close
DERA, Farnborough
August 2000

[1] www.titaniuminfogroup.co.uk

New titanium applications in the UK – an overview

M WARD-CLOSE
Defence Evaluation and Research Agency (DERA), Farnborough, UK

ABSTRACT

This paper describes recent developments in the application and use of titanium in the UK. Continuing evolution and consolidation within the titanium industry is discussed. Examples of new applications are described from the UK aerospace industries, and other non-aerospace examples including ground based military hardware, off-shore equipment and architectural cladding. It is anticipated that as titanium alloys continue to expand in use and diversify into non-aerospace sectors, supply will become more stable and the historically cyclical nature of the titanium market will disappear. Prospects for reduction in the cost of titanium are discussed and several new initiatives in this area have the potential to substantially reduce the cost of titanium, possibly, for non-aerospace grade titanium, to half the current price or less.

Key words: titanium, welding, friction welding, aeroengines, aerospace, metal matrix composites, ordnance, architecture.

1 INTRODUCTION

The process of consolidation within the European titanium industry has continued, with the take over of IMI Titanium in Birmingham by the US company Timet, to form Timet UK. As part of their infrastructure development Timet UK have recently installed a Radial Forge Machine (RUMX) capable of deep section deformation in two axes and rapid breakdown of ingots to billet and bar. RTI Titanium have also expanded their operations in the UK with a new distribution centre near Birmingham, and new milling, water jet cutting and CMM equipment for the supply of sheet, plate and bar.

Two themes emerge in this overview of titanium applications in the UK. Firstly, a general increase in non-aerospace activity and a marked increase in the level of interest in titanium from all market sectors, ranging from sporting goods to medical equipment and from architecture to off-shore. Secondly, it is clear that the UK has established a leading position in titanium superplastic forming and diffusion bonding (SPFDB) and it is noticeable how many different applications using this technology are now reaching maturity.

Although it is the intention of this overview to concentrate on industrial applications of titanium, it must be acknowledged that much excellent titanium research work is being carried out in UK universities and research institutions. In particular at Imperial College, Birmingham IRC, Oxford, Cambridge and Leeds Universities.

2 JOINING

TWI (The Welding Institute) have continued to develop friction welding technology for titanium. Figure 1 shows a demonstration of rotary friction welding of a 240 mm diameter Ti-6-4 pipe, with a 16 mm wall thickness. A high integrity weld is formed in only 10 s, whereas a skilled welder would take about 8 hrs to produce a similar joint using TIG. TWI have also demonstrated friction stir welding of titanium. This low distortion/non-melting process, which is now well established for aluminium alloys, involves forcing a special tool along the line of the joint, producing a highly deformed region which has excellent integrity and strength. However, friction stir welding is still at an early stage of development for titanium. Much work remains to be done on tool life, joint quality and process optimisation before it can be applied commercially.

Titanium is under consideration for light weight fuel cells for electrical generation. At DERA, diffusion bonding has been used to produce highly complex multi-layer titanium fuel cell plates. Research is driven by the need for light-weight, compact fuel cells for electrical powder generation in weapon systems, vehicles and man-portable military equipment. In the present development, titanium replaces much bulkier carbon plates and the intricate internal channels provide separate paths for hydrogen, oxygen and cooling medium. Figure 3 shows a rectangular, five layer fuel cell plate and Figure 4 is a circular design where the centre gas entry gives a more even loading of the structure.

3 AEROSPACE APPLICATIONS

The introduction of the Rolls-Royce Trent 700 aeroengine for the Airbus A330 aircraft and the Trent 800 for the Boeing 777 has seen continuing advances in the use of titanium alloys in engine technology. In a new development for Rolls-Royce both engines use Ti-6-2-4-6 compressor discs, optimised for strength and toughness using beta and trans beta-forging.

The design of the Rolls-Royce Trent hollow wide-chord fan blades has continued to evolve. Earlier designs used a brazed titanium honeycomb construction, but the current blades use superplastic forming and diffusion bonding (SPFDB) to produce light weight internal stiffening in a one piece construction. For the latest addition to the Trent family, the Trent 8105, a 105,000 lb thrust engine intended for future 'superjumbos', Rolls-Royce have developed a scimitar shaped swept blade for enhanced aerodynamic performance and improved foreign object damage (FOD) resistance.

Rolls–Royce continue to develop their blade and disc technology for future generations of engines. Figure 4 shows an advanced bladed disc (BLISC) which has substantial potential to reduce engine weight. Instead of the usual slotted blade and disc arrangement the blades are integrally bonded to the disc using linear friction welding.

Linear friction welding allows for precise alignment of the blade, combined with a very narrow heat affected zone and minimum distortion. During the welding operation contaminated surface material is extruded from the interface region and is removed in the final machining. This technique will also be used to replace blades. Damaged or life expired blades will be removed from the disc by electro-discharge machining (EDM) and the new blade joined by linear friction welding; a process that can be repeated several times.

Looking even further ahead, work is continuing on the fibre reinforced bladed ring (BLING). Here the ring is reinforced with hoop wound silicon carbide fibre metal matrix composite (MMC). In future engine designs BLINGS will be welded together to form a complete compressor drum, saving up to 70% of the rotating mass compared with a conventional bladed disc and also opening a central cavity that can be utilised for gas flow and enhanced aerodynamic performance. In a joint programme, DERA and Rolls-Royce have demonstrated the fabrication of BLINGS using the foil-fibre (F-F) diffusion bonding method for MMC fabrication and also the newer matrix coated fibre (MCF) method. In the MCF route SiC fibres are pre-coated with titanium alloy by electron beam evaporation and the coated fibres are then diffusion bonded and consolidated into the MMC component. A new, high capacity, fibre coating facility has been established at DERA, Farnborough and fibre is now routinely coated with Ti-6-4 or Ti-6-2-4-2 in lengths up to 2000 m.

British Aerospace have made extensive use of titanium in their new joint European fighter aircraft. The airframe of Eurofighter (Typhoon) has approximately 12% by mass of titanium alloy products. The majority of this mass is accounted for by the major frames and longerons within the rear fuselage area, around the engines, and also the wing and fin attachments brackets. Most of these items are machined from forgings and some of the brackets utilise Electron Beam welding in their manufacture. The remaining usage of titanium is in sheet applications with more than half being processed using SPF or SPF/DB techniques. The foreplanes are manufactured using a four sheet X-core structure, this route being selected as the most cost effective solution to the stiffness and flutter requirements of this part. Around the keel area the keel structure and tunnel skins are all manufactured using SPF or SPF/DB techniques. There is currently only one titanium casting on Eurofighter (the rudder torsion fitting) but BAe are currently actively pursuing the increased use of titanium castings for future projects.

4 NON-AEROSPACE

Titanium alloys are continuing to make significant penetration into non-aerospace markets such as bio-medical, marine, architectural, armoured vehicles, ordnance, automotive, sports and leisure and many others. The following are just a few examples from the more technically innovative new applications in the UK.

Rolls-Laval have introduced a new line of compact high-integrity heat exchangers of the plate-fin type and intended for high pressure applications in the hydrocarbon and chemical industries. Made entirely of titanium, the design features SPFDB of the plate element to give high integrity bonds with no weld bead, no flux and no heat affected zone. In a typical gas cooler for North Sea operation, the new design is seven times lighter than a conventional shell-and-tube heat exchanger of the same capacity.

In a collaborative programme between DERA and a number of UK companies, a new toroidal shaped titanium air bottle has been developed which is lighter than existing steel bottles and has a number of other advantages compared to conventional cylindrical bottles. The production process involves superplastic forming in two pieces, welding and then over wrapping with aramid fibre. The fibre is wound dry, and under high tension, using a specially developed toroidal winding machine.

For divers or fire fighters the toroidal air bottle provides a more convenient shape, less likely to snag on obstructions and carried low on the back is more comfortable to wear. A further added benefit is that in the centre hole the valve gear is protected from damage and can therefore be of lighter construction. Table 1 shows a weight comparison between a titanium/aramid toroidal 207 bar, 9 litre, bottle, steel and aluminium toroidal bottles made by the same process, and conventional cylinders in steel, plain aluminium and carbon fibre overwrapped aluminium. At 5 kg the new titanium toroidal bottle is slightly heavier than an aluminium/carbon fibre cylinder, but does not require the back plate normally used for cylinders and has all the ergonomic advantages described above.

Table 1. Toroidal air bottle – weight comparisons.

	Toroid Aramid overwrap	Conventional cylinder	Carbon fibre overwrapped cylinder
Titanium alloy	5	-	-
Steel	6	12	-
Aluminium alloy	6	14	5 (incl. 1 kg back plate)

Titanium alloy is now under serious consideration by the UK MOD for land based military equipment. With the modern emphasis on light-weight air mobile forces titanium is attractive for weight reduction. For vehicle armour, titanium has nearly the same ballistic performance as the current rolled steel armour (RSA) with approximately 40% lower weight. However, cost remains a major concern.

In an example of a titanium application in ordnance, a UK company, BAE SYSTEMS has introduced the Ultralightweight Field Howitzer which is approximately 60% titanium by weight. At 3,745 kg the new gun is less than half the weight of the existing 155 mm howitzer used by the US army, and is fabricated from Ti-6Al-4V mainly in the form of welded plate and sheet. The initial order is to supply 10 guns for evaluation by the US marines, and if successful this is likely to lead to orders for over 1000 guns from the US Marines and the US Army.

In the architectural field, despite a higher cost than stainless steel, titanium is finding favour both as external cladding and for internal decorative work, where the soft grey appearance of titanium is preferred to the bright grey of stainless steel. The Glasgow Science Centre, the UK's first titanium clad building is due to open in the year 2000. Designed by the Building Design Partnership, the building will be clad with 6000 m^2 of 0.3 mm thick titanium sheet.

The wide range of colours possible with anodising makes titanium attractive for jewellery and other decorative arts. Susan Vernon at the University of Derby, School of Art and Design has produced a range of unusual titanium cutlery, including the camping tools and ice cream spoons.

5 CONCLUDING REMARKS

Both the volume and diversity of titanium alloy usage is continuing to expand in the UK, and organisations such as the Titanium Information Group (TIG)[1] in the UK and the International Titanium Association (ITA) in the US continue to do an excellent job publicising titanium and disseminating information on the use of titanium. It is clear that as awareness of titanium increases generally there is a huge level of interest by architects, designers and engineers in using titanium, principally for its low weight compared to steel and its excellent corrosion resistance. However, cost remains a major concern and it is to be hoped that the various initiatives both in Europe and the US aimed at reducing the cost of titanium alloys will bear fruit in the near future. For example, titanium is an excellent lightweight substitute for steel tank armour, but at present the cost is reckoned to be about twice the level needed for it to be seriously considered for this high volume application. One option for cost reduction is to utilise new cold hearth melting technology as an alternative to the traditional vacuum arc melting (VAR). Open hearth melting may allow single melting of ingots, instead of the normal double or triple melting required for VAR and in the USA evaluation of experimental e-beam single melt Ti-6Al-4V is showing promising results. It is anticipated that this approach may allow price reduction for non-aerospace grade titanium alloy to below £10/kg.

6 ACKNOWLEDGEMENTS

The author is indebted to the following people for their help in preparing this overview paper: Phil Threadgill (TWI), Dave Mayo (BAe), Phil Doorbar (RR), Dave Rugg (RR), John Fowler (Rolls-Laval), Barry Lakeman (DERA), John Cooke (DERA), Andy Wisbey (DERA), Susan Vernon (Derby University, School of Art and Design), David Peacock, Roger Thomas (Timet), Bob Ringrose (RTI) and Alistair Elder (BDP).

[1] TIG can be reached at www.titaniuminfogroup.co.uk and ITA at www.titanium.org

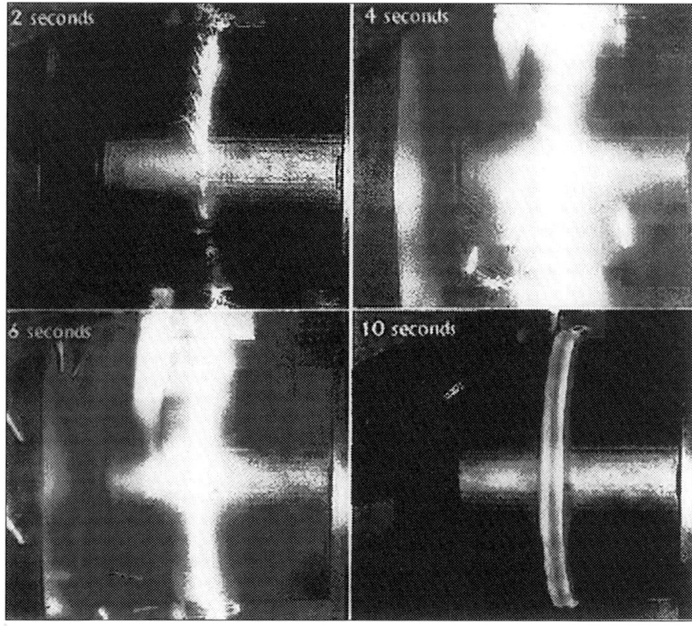

Figure 1. Rotary Friction Welding of a 240 mm diameter Ti-6-4 pipe, with a 16 mm wall thickness – a skilled TIG welder would take approximately 8 hours to complete a similar weld.

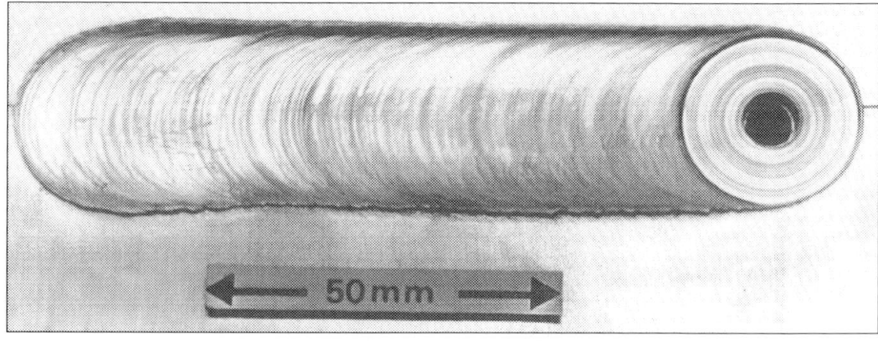

Figure 2. Experimental friction-stir weld in Ti-6Al-4V.

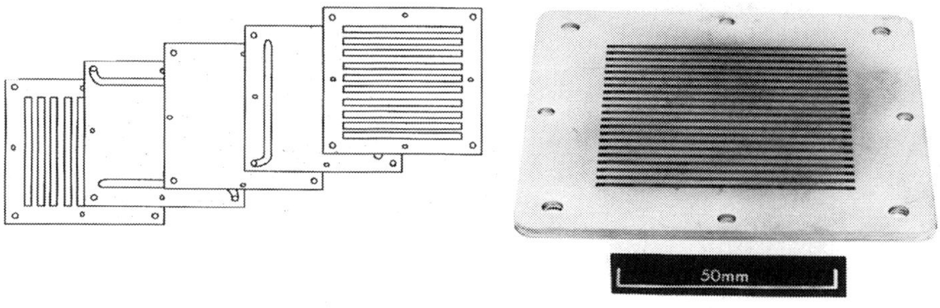

Figure 3. Titanium diffusion bonded 5 layer fuel cell plate.

Figure 4. Circular diffusion bonded fuel cell plate, showing centre gas entry for better stress distribution.

Figure 5. Rolls-Royce Trent 700 and Trent 800 aeroengines, showing titanium SPFDB wide-chord fan blades and SPF spinner farings.

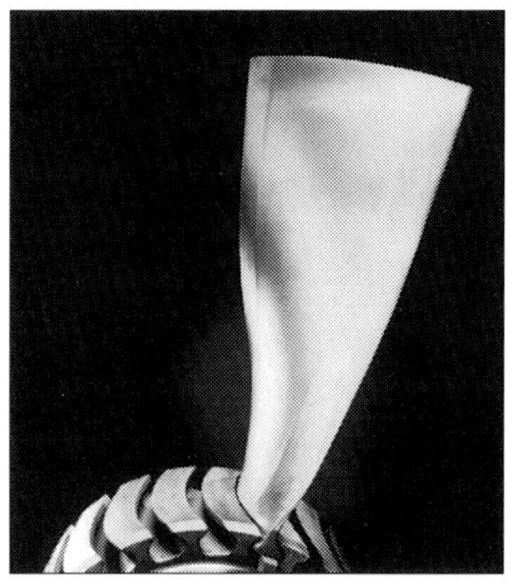

Figure 6. Rolls-Royce SPFDB Wide Chord Fan Blade

Figure 7 Rolls-Royce SPFDB Swept Wide Chord Fan Blade

Figure 8. Rolls-Royce Advanced Titanium BLISK

Figure 9. Titanium Ti-6Al-4V linear friction weld.

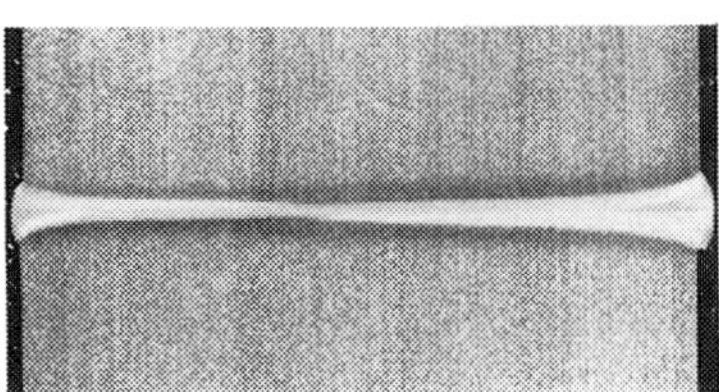

Figure 10. Titanium Ti-6Al-4V linear friction weld microstructure

Figure 11. Fibre reinforced bladed ring (BLING)

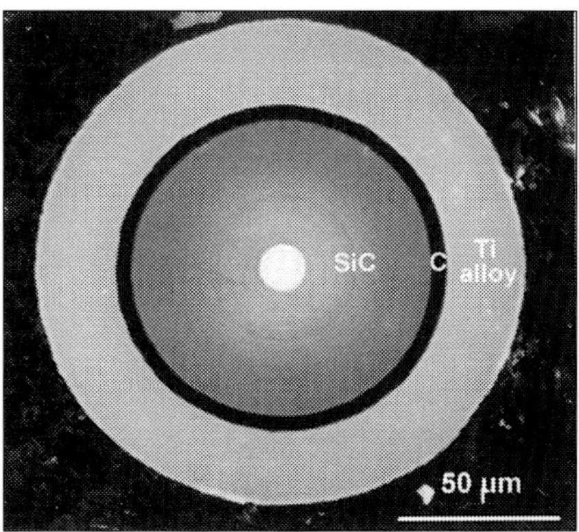

Figure 12. DERA Matrix Coated SiC Fibre for Ti MMCs

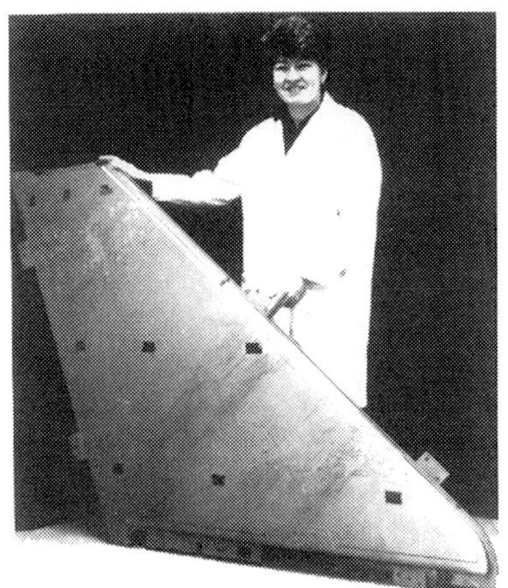

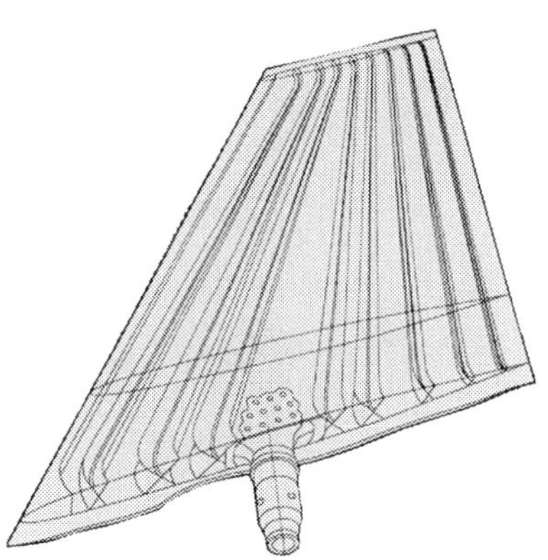

Figure 13. BAe Typhoon SPFDB Foreplane.

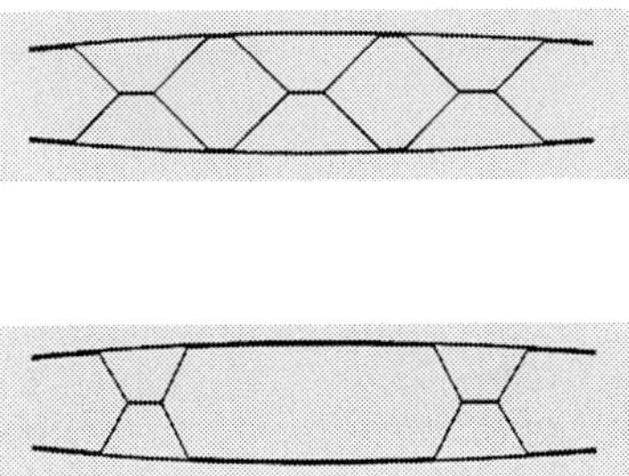

Figure 14. Cross section of BAe Typhoon 4 sheet SPFDB titanium foreplane, showing evolution of design from early concept (top) to current low mass design (bottom).

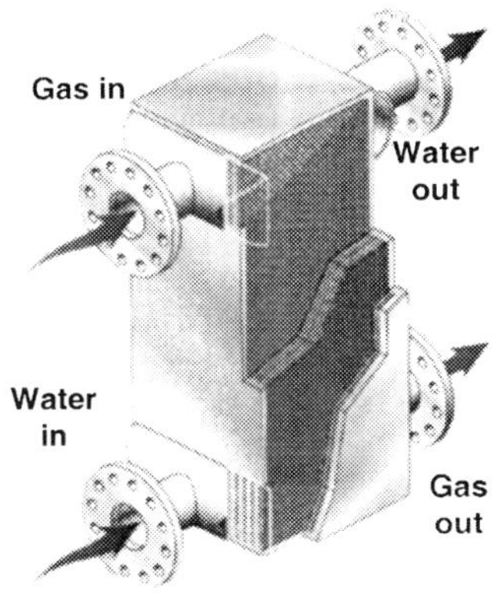

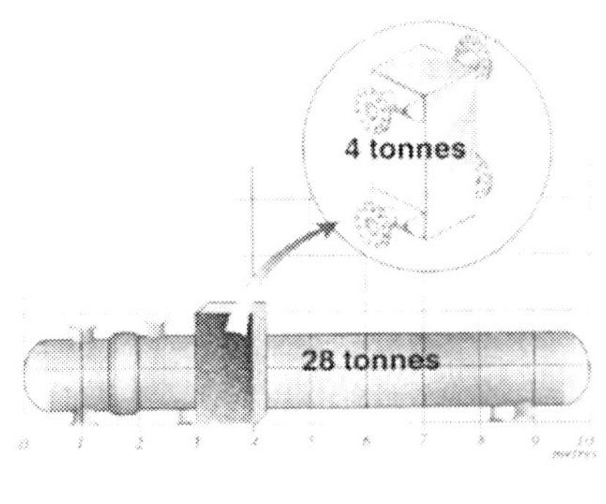

Figure 15. Rolls-Laval SPFDB titanium heat exchanger.

Figure 16 Comparison with conventional shell-and-tube heat exchanger of the same capacity (weight when full).

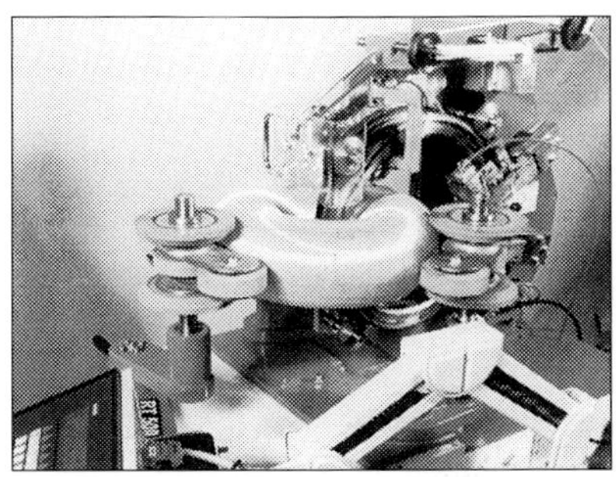

Figure 17. Special toroidal winding machine for overwrapped titanium air bottle.

Figure 18. Titanium toroidal air bottle compared with conventional cylinders.

Figure 19. VSEL Ultralightweight Field Howitzer – 60% Titanium Alloy.

Figure 20. Glasgow Science Centre – 6000 m^2 titanium roof.

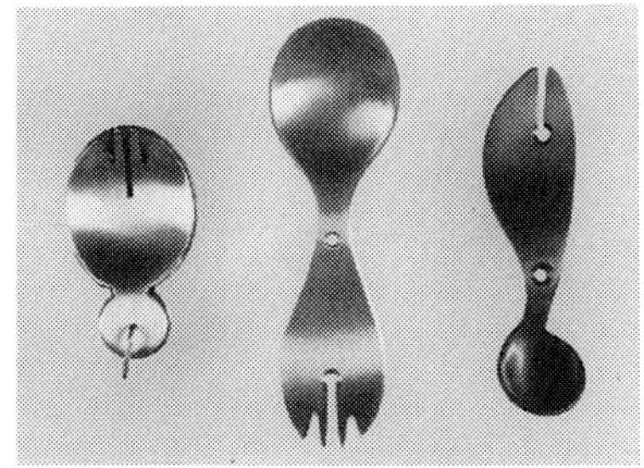
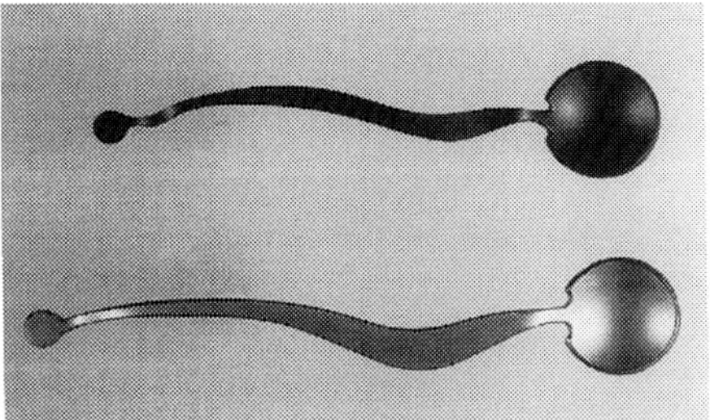

Figure 21. Titanium camping cutlery and ice cream spoons.

The recent advancements in bone and joint implant technology

P S UNWIN
Stanmore Implants Worldwide Limited, Centre for Biomedical Engineering, Royal National Orthopaedic Trust, Stanmore, UK

Joint replacement surgery is one of the most successful surgical procedures transforming the lives of patients crippled with joint pain. The demands placed on the joint replacement implant are extreme, requiring high performance materials and with today's economic pressures, cost-effective production.

WHAT IS AN IMPLANT?

An implant is a foreign material placed in the body to replace or aid defective organs or tissues. Over the last five decades implant technology has boomed and today there is a large array of materials used. These materials range from metals to ceramics to polymeric materials and more recently reengineered biological tissues. This paper looks at just one metal, Titanium Alloy and it's use in bone and joint replacement in orthopaedics. Orthopaedics is the branch of medicine that focuses on the restoration of the function of the skeletal system. The word orthopaedics is derived from Greek meaning straight (ortho) child (pais), joint replacement. Joint replacement or to give it's proper name – arthroplasty again is derived from Greek and means joint (arthron) forming (plassein).

BONE AND JOINT REPLACEMENT

Bone and joint replacement is not new, the Egyptians were known to have attempted bone replacement. A mummy discovered recently was identified to have femur (thigh bone) replaced with a piece of wood. However, there has been little development in bone and joint replacement until 1891 when Theophilus Gluck replaced a femoral head with a piece of ivory. A few years later a thin layer of gold was inter-positioned between the femoral head and the hip joint socket. The malleability of gold permitted the surgeon to form the gold over the head during surgery. Both of these solutions were aimed at eliminating pain emanating from the joint due to joint disease. Unfortunately both solutions were not successful but the principal indication for joint replacement is osteoarthritis and that was much less prevalent at the turn of the century. During the 20th Century there has been a significant increase in life expectancy and population. Little progress in joint replacement was made until the 1940's when metallic hips were introduced. The original hip replacement of the 1940's and the 1950's used a metal-on- metal coupling. During the 1950's metal-on-metal hip replacements were originally used, these were fabricated from stainless steel, although this was soon superseded by cobalt chrome molybdenum alloy. Cobalt chrome molybdenum alloy had the advantage of reducing friction; it was also recognised that it was advantageous to use the tribiological pair to be of identical materials to avoid the galvanic corrosion.

The most significant breakthrough came in the 1960's with Sir John Charnley's introduction of the cemented ultra low friction arthroplasty. By designing a small diameter head integral with the femoral hip stem that articulated with an Ultra High Molecular Weight Polyethylene (UHMWP) acetabular cup (that fits into the hip socket) resulted in an ideal combination that has dominated orthopaedics since. The demise of metal-on-metal joint replacements was a result of a number of inter-related factors including, friction, precision manufacturing of the couple and the concerns over the biocompatibility. Studies undertaken in the 1960's raised the serious issue of carcinogenicity as a result of metal ion release from the metal on metal joint replacement. Studies undertaken in rats indicated the increased risk of forming cancers due to the toxicity of the constituents of the metal alloy. Prior to 1962 implants were press-fit into the recipient bone. The introduction of polymethylmethacrylate (PMMA), that was being used successfully in dentistry, secured the implants stems into the bone mechanically. In orthopaedics PMMA is incorrectly termed as bone cement, currently it is used in approximately 50-60% of joint replacements. In terms of technology advancement the introduction of polyethylene and PMMA revolutionised joint replacement. There has been little change in the design of the Charnley hip replacement and today it is still a popular device.

The success of the surgical joint replacement procedure has resulted in a huge demand coupled with the average age of the patients lowering from upper 70s to the 60s. The younger patients are more demanding of the device and will live for another 20 or 30 years. This increase in the long-term durability of the implant has led orthopaedic surgeons and scientists to seek high performance materials. Over the last two decades the long-term performance of both PMMA and polyethylene has been questioned.

The progressive wear of polyethylene acetabular cups of approximately 0.1mm per year limits the life of the cup to approximately 10-15 years, thus in the younger patients revision of the cup is normally inevitable. The polyethylene wear debris migrates to the implant-bone interface and is linked to a complex multi-factorial process of implant loosening in the host bone. Furthermore the long-term mechanical degradation of PMMA exacerbates the loosening process. Integration of the metallic implant directly to the host bone eliminates the need for the "bone cement" layer. Osseointegration (bone in-growth) can be encouraged by a variety of techniques used including roughening the implant surface, sintering layers of small metallic beads, or plasma coating the implant surface with thin layer of metal (such as titanium alloy) or ceramic (typically hydroxyapatite).

In the UK today, over 50,000 hip replacements are undertaken annually, in addition there is approximately 30,000 knee replacements and an assortment of other joint replacement including Shoulder, Elbow and Ankle.

MATERIALS USED IN JOINT REPLACEMENT

The three principal metals used for joint replacement are stainless steel (in various forms), Cobalt Chrome Molybdenum and Titanium Alloy. In addition, ultra high polyethylene is routinely used and there are also ceramics, alumina (AlO_3) and zirconia (ZrO). All three have been used over the last four or five decades. Titanium has been used since the 1960's, there are a number of specific advantages of using titanium alloy, however there are some disadvantages, but in the leading edge of custom implants the titanium alloy is the material of choice. The commonest alloy of titanium used for orthopaedic implants has been Ti-6Al-4V (Ti 318). Over the last decade other titanium alloys have been introduced and included Ti-6Al-7Nb and Ti-5Al-2.5Fe. The drive for the introduction was primarily because of the potential cytotoxicity of vanadium. Recently, second-generation titanium alloys have been developed and these offer lower modulus and superior biocompatibility. Long has provided a comprehensive review of titanium in orthopaedics (Long, 1988).

Over the last 4 decades, titanium alloy form 318(Ti-6 Al-4V) has been used for various implants throughout the body (Figure 1). Titanium alloys favourable properties are it's high strength, relatively low modulus of elasticity, it's resistance to corrosion, excellent biocompatibility and for certain situations such as segmental bone replacement it's lightness, however it's disadvantage primarily is it's wear resistance.

The thermo-mechanical processing of titanium alloys has allowed for the production of high strength materials. The demands placed on human joints are considerable, it is estimated that subjects take approximately 1.8 million walking steps per year. A study of human activity identified that hip replacement subjects undertake less activities compared to non impaired subjects, but the resultant forces at the hip joint are of similar magnitude. Whilst walking it was calculated that approximately three times body weight is transmitted through the hip joint, exiting from a car this can increase up to five times body weight. Common activities, sitting, walking and standing are relatively consistent between the two sets of patients although the transfer from sitting to standing and stair climbing were rarer with patients who had undergone hip replacement, however it is often these rare activities that are important because they represent overload events which influence the fatigue life of the implant. Titanium alloy is considered to be excellent at resisting fatigue, but it recognised that it can be extremely sensitive to prior thermo-mechanical processes and subsequent processing techniques.

The fatigue life can be enhanced by cold working the surface using shot-peening techniques that increase the compressive residual stresses and reduces fatigue failure by limiting crack initiation. Increasing the surface roughness through grit blasting has been identified to reduce the fatigue life, this leads to early crack initiation. The initiation and growth of fatigue cracks in titanium alloys are controlled by complex microstructural and mircomechanical effects. Shot-peening and grit blasting are used routinely to roughen the implant surface to permit either osseointegration or as surface preparation prior to ceramic coating. The complex interaction between pre-treatment (work hardening and secondary manufacturing processing) and surface coatings does require detailed investigation and there is an increasing need for standard test procedures to ensure the safety of devices.

The modulus of elasticity is a very important parameter, and titanium alloy in comparison to the other two metals is lower and preferable although it is considerably stiffer than bone

(Table 1). The mismatch of the moduli results in stress shielding of the bone leading to a decrease of bone thickness and bone mass. Bone (a complex anisostropic composite that is a living organ) adapts to fulfil it's mechanical function, it adapts to the loads and stresses placed upon it. Wolf identified the control mechanism that is now referred to as Wolf's Law, increasing the loads placed upon bone will elicit an increase in bone mass and thickness. Increasing the stiffness by the insertion of a metal stem into the shaft bone leads to stress shielding. It has been considered that reducing the stiffness of the implant material to 40-50 GPa (twice that of cortical bone) may help to reduce stress yielding and a number of attempts have been tried using carbon composites. A number of new titanium alloys are being developed with lower moduli and are providing an exciting prospect for arthroplasty.

Table 1. The mechanical properties of the 3 most commonly used orthopaedic alloys compared with cortical bone.

Metal Alloy	Youngs Modulus (*E* GPa)	UTS (MPa)	Yield Strength (MPa)
Ti-6Al-4V	110	960-970	850-900
Stainless Steel 316L	200	465-950	170-750
Cobalt Chrome Molybdenum	220	600-1795	275-1585
Cortical Bone	25-30	90-140	-

Titanium alloy in orthopaedics has long been recognised as being resistant to corrosion, it has been known that titanium alloy has four times the corrosion resistance of medical grade stainless steel. Many of the current joint replacements are modular, providing the surgeon with increased choice of sized components at the time of the surgical procedure. However the taper junctions are prone to fretting corrosion and crevice corrosion. Studies of retrieved titanium alloy femoral stems with cobalt chrome alloy femoral heads have shown some evidence of corrosion (Willert, 1996) & Jacobs 1998). It has been considered that this has been a result of micro-motion on the taper connection and the micro-motion deforms the passive oxide film. This formation of the passive oxide film increases the negativity and becomes more oxidised and continues the process of fracture and re-passivation of the oxide layer at each of the cyclic movements. The reduction of oxygen in the system also decreases the pH at the micro-crevice. Corrosion has been identified on implants that have been scratched or that have scratched surfaces. Scratches have occurred during production, during transit, in preparation for surgical procedure or whilst inserting or post operatively by interfacial micromotion or third body wear. It must be stressed however that the extent of corrosion at these interfaces is significantly less than other materials particularly those of the stainless steel variety.

The wear of titanium has been well documented. Titanium nitriding and ion implantation have been used to improve the surface hardness and studies are reporting encouraging results (Pappas, 1995). In relation to wear there are concerns not only of the mechanical integrity of the device, but also of generation and distribution of the wear particles. The reactions range from local inflammatory to allergic reaction and with a potential of a carcinogenetic reaction.

The extent of the local tissue reaction is determined by the volume, size and shape of the titanium particles (Blunn, 1991). Small round particles are engulfed by cells known as macrophages whereas large particles are encapsulated by a number of cells (known as a foreign body giant cell). The extent of the tissue reaction where there was minimal volumetric wear results in particles being found in fibroblastic cells. Increasing the volumetric wear there was a histiocytic reaction, where the particles of titanium were engulfed by these scavaging macrophages. In tissues that contained large volumes of titanium there was an infiltration of lymphocytes, this suggests that there is a immuniological response and where there is very high volumes of wear tissue necrosis occurs and these are infiltrated by a high number of lymphocytes and macrophages. Titanium wear debris has also been found in distant tissues, there are a number of reports that have identified titanium debris in the body's lymph nodes, in particular the inguinal (groin) lymph nodes. Other studies have shown there is metal release and excretion of titanium ions. The systemic toxicity of the ions and by products of the corrosion and wear processes are as yet unknown.

Custom implants – use of titanium alloy

Currently, the vast majority of joint replacements are mass produced. Casting and forging processes are used to produce implants in a range of sizes to accommodate variability between patients. For the majority of patients "off-the-shelf" joint replacements provide a satisfactory solution. However, occasionally "off-the-shelf" replacements are just not suitable for one or more reasons including abnormal bone geometry, or severe bony destruction. The alternative is to use a custom-made patient specific implant. The Centre for Biomedical Engineering, located at the Royal National Orthopaedic Hospital, Stanmore has been researching, designing and manufacturing custom-made joint and bone replacements since 1949. Currently over 6,000 of these patient specific implants have been used and range from miniature hip and knee replacements used in dwarfism to the replacement of complete bone such as the femur or humerus for conditions such as osteosarcoma, a malignant bone tumour (cancer).

The majority of these 6,000 implants have been to replace a large segment of bone (the joint plus some or all of the shaft bone). These bone replacement implants are known as massive replacements and are typically used when there is a bone tumour or extensive bony destruction associated with a failed joint replacement. Unfortunately bone tumours occur at a relatively young age and some of the patients requiring limb-salvage (shaft bone replacement) are skeletally immature. These children (the youngest patient undergoing bone replacement was 2 years 9 months at the time of the operation) have designed special implants that are extendible to enable limb length equality to be maintained (Plates 1 & 2). The worm drive extending mechanism is contained within the implant shaft and is extended using a key that is passed through a small incision in the skin. In addition to the needs of an extending mechanism, implants for children are frequently scaled down and the stems that secure the implant in the remaining bone shaft may be under 7 mm in diameter. The demands placed on the implants by these active children are significant. The surgical options for malignant tumour occurring in a limb bone are few and include amputation or replacement of the shaft bone either with bone from a bone bank (allograft) or transfer of the patients own bone (such as the fibulae) or metallic implant. In the UK, the predominant surgical option is the use of a metal implant and the long-term results of these devices are very encouraging. There are now a number of patients with proximal femoral or distal femoral replacements who under went limb-salvage surgery over 20 years ago. However, aseptic loosening of the implant's intra-medullary stem in the remaining segment of bone was a major cause of implant related failure, particularly in the skeletally immature patients. The original massive replacements were secured into the bone using PMMA, but using technology taken from joint arthroplasty, uncemented (press-fit) hydroxyapatite coated intra-medullary stemmed massive implants have reduced the incidence of aseptic loosening considerably. Reducing the rate of failure and subsequent revision surgery has a tremendous impact on patient quality of life and also hospital budgets.

Since 1962 titanium alloy (Ti 318) has been the material of choice for the bulk of the implant. In addition to high strength, corrosion resistance, lightness and biocompatibility, titanium alloy can be machined by skilled technicians with relative ease. In the treatment of bone tumour the timing of the surgical procedure is often critical and therefore there is a need to design and fabricate the implants rapidly. Using modern manufacturing resource planning techniques, custom-made implants are fabricated from sub-assembled modularised components, thus reducing the amount of machining required. The manufacturing time varies enormously depending on the complexity of the device. Straightforward devices such as proximal femoral replacements that comprise of 4 components (a femoral head, trochanteric neck, shaft and stem) can be fabricated within 2-3 hours. Complex implants such as extra-small extendible distal femoral and knee replacements may take in excess of 30 hours and hemi-pelvic replacements machined from one piece of titanium alloy bar (150 mm diameter) have exceeded 50 hours of fabrication.

In addition to the massive replacements, the other patient specific implants are joint replacements, but typically are for complex reconstructive scenarios such as abnormal bone geometry or a revision of a failed standard joint replacement. Some of the implants requested are for patients with dwarfism and again miniaturised implants are required.

One of the distinct advantages of having a patient specific implant is that the implant is designed to fit the patient in contrast to "off-the-shelf" implants where the patient is fitted to the implant. The close fit of the implant in the host bone provides a more even distribution of the transfer stresses and significantly reduces micro-motion. Optimal stress transfer coupled with minimal micro-motion enables osseointegration of the implant to take place leading to long-term implant survival.

During the last two decades, there has been a dramatic expansion in the development of Computer Technology in a wide variety of industries, this has included medicine and in particular has been beneficial to the advancements of orthopaedic implants. The orthopaedic industry has transferred over some sophisticated technologies including that of imaging, machining and modelling and together these have provided the capabilities of producing some of the most complex joint replacements.

During the 1980's there was a dramatic expansion of sophisticated computer aided design and manufacturing technologies. The Centre for Biomedical Engineering at Stanmore developed a unique hip design workstation for the design of hip replacements using bi-planar radiographs. The hip design workstation comprised of a digitising system, computer generated models from which the hips are designed and then the hip design information was processed for the machining program, this being downloaded to a computer controlled milling machine. All the necessary software for mapping the proximal femoral canal, designing of the femoral stem and machining implants from titanium alloy bar was developed in-house, this allowed us to produce three dimensional models from digitised bi-planar radiographs to provide a method of machining the components rapidly and precisely.

The first CAD-CAM femoral hip stem was designed, fabricated and inserted in 1989, currently over 1,350 of these CAD-CAM Hip stems have been inserted, the majority of which have been for complex primary reconstructions of the hip joint or revisions of failed primary joint replacements (Plates 3 & 4). The key to these replacements is the precision of the design leading to the optimal fit and fill of the proximal femoral canal. The stems are uncemented and require to be in close proximity to the adjacent bone. Various design features can be incorporated into the actual design including a collar and flutes on the distal section of the stem that aid rotational stability. The combination of the hip design software and the computer controlled milling machine provides a very cost-effective manufacturing route. CAD-CAM hip replacements can be milled within 1-2 hours depending on the actual dimensions of the device. Once the milling is complete, the femoral stem is polished and the femoral neck taper is precisely turned to accept either a cobalt chrome alloy or ceramic femoral head. The Centre for Biomedical Engineering has worked on methods to evaluate the clinical performance of these devices and studies of axial migration of the stem replacement has shown that there is only 1.5mm of axial migration after two years of implantation. This is very encouraging as it demonstrates that the stability of the device in the canal is great which encourages osseointegration of the hydroxyapatite coatings on the implant with the adjacent bone.

Recently, there has been a number of patients who have been referred to the Centre for joint replacement that have demanded usage of some very sophisticated imaging and modelling technologies. Severe deformities of the hip replacement make the visualisation of the femoral canal space and the joint alignment difficult to visualise. Using CT data, the bone can be modelled and this data can be sent to a bureau to produce a stereolithography to produce a resin model. Working in collaboration with the surgeons using this model, an implant can be designed that meets the requirement of the patient. The realistic model provides the surgeon with the ability to study the deformity, helping the engineers design the implant and in combination even carrying out trial surgery. Using this rapid prototyping technology we have been able to successfully provide surgeons with titanium alloy hip replacements that until recently could not have been provided.

These technologies using computer aided design and computer aided manufacturing have also been transferred to the area of knee replacements and the automation of the customised knee replacement requiring only the input of anatomical dimensions derived from patient radiographs is now a possibility. We have provided superstabiliser knee replacements for a number of patients who have dwarfism. The surgeon has been unable to use off the shelf or standard joint replacements because of the size of the patient's knee, the only solution was to have a custom-made knee replacement.

With the major breakthroughs in genetic engineering there are a number of recent advances in biological reconstruction methods for degenerative joint diseases and it is likely that metal joint and bone replacements will still be in use for the foreseeable future.

FURTHER READING

1. Clerc C. *et al.* Assessment of Wrought ASTM F 1058 Cobalt Alloy Properties for Permanent Surgical Implants. *Journal of Biomedical Materials Research 1997 V. 138 N.3*
2. Long M, Rack H.J. Titanium alloys in total joint replacement – a materials science perspective. *Biomaterials 19 (1998) 1621-1639.*

3. Cales B, Stefani Y. Risks and Advantages in Standardization of Bores and Cones for Heads in Modular Hip Prostheses. *Journal of Biomedical Materials Research 1998 Vol. 43 No 1.*
4. Verdonschot N, Huiskes R. Surface roughness of debonded straight-tapered stems in cemented THA reduces subsidence but not cement damage. *Biomaterials 19 (1998) 1773-1779.*
5. De Mol B A.J.M, Van Gaalen G L. The Editor's Corner: Biomaterials Crisis in the Medical Device Industry: Is Litigation the Only Cause? *Journal of Biomedical Materials Research 1996 Vol.33 No.1.*
6. Black J. Properties of Natural Materials. Orthopaedic Biomaterials in Research and Practice.
7. Dobbs H S, Robertson J L M. Alloys for orthopaedic implant use. *Engineering in Medicine 1982. Vol 11 N0.4.*
8. Semlitsch M, Weber H, Steger W. 15 Years' Experience with Ti-6Al-7Nb Alloy for Joint Replacements. *Titanium '95: Science and Technology.*
9. G W Blunn *et al.* The Reaction of tissues to Titanium wear generated from massive segmental bone defect prostheses. *Complications of limb salvage, Prevention, Management and Outcome (ISOLS Montreal) VIII-Endoprosthetic Design. 1991.*
10. Lambert R.D, Anthony M. E. Standardization in Orthopaedics. The Growth and Activities of ASTM's Arthroplasty Subcommittee. *ASTM Standardization News, August 1995.*
11. Pappas M, Makris G, Buechel F.F. Titanium Nitride Ceramic Film Against Polyethylene. *Clinical Orthopaedics August 1995 Vol. 317.*
12. Vince D.G, Hunt J. A, Williams D F. Quantitative assessment of the tissue response to implanted biomaterials. *Biomaterials 1991, Vol 12 October.*
13. Shinto Y *et al.* Inguinal lymphadenopathy due to metal release from a prosthesis. *The Journal of Bone and Joint Surgery Vol 75-B No.2 March 1993.*
14. Cordero J, Munuera L, Folgueria M.D. Influence of metal implants on infection. *The Journal of Bone and Joint Surgery Vol 76-A No. 5 September 1994.*
15. Kornu R *et al.* Osteoblast Adhesion to Orthopaedic Implant Alloys: Effects of Cell Adhesion Molecules and Diamond-Like Carbon Coating. *The Journal of Orthopaedic Research Vol 14 No.6 1996.*
16. Mjöbery B, Theories of wear and loosening in hip prostheses Wear-induced loosening vs loosening-induced wear – a review. *Acta Orthop Scand 1994; 65 (3): 361-371*
17. Black J. Biological Performance of Materials. Fundamentals of Biocompatibility. Second Edition. Marcel Dekker Inc 1992. USA.
18. Jacobs J. *et al.* Corrosion of metal orthopaedic implants. *The Journal of Bone and Joint Surgery Vol 80-A No2, February 1998..*
19. Willert H-G *et al.* Crevice corrosion of cemented titanium alloy stems in total hip replacement. *Clincal Orthopaedics and Related Research. Vol 333: 51-75. 1996.*

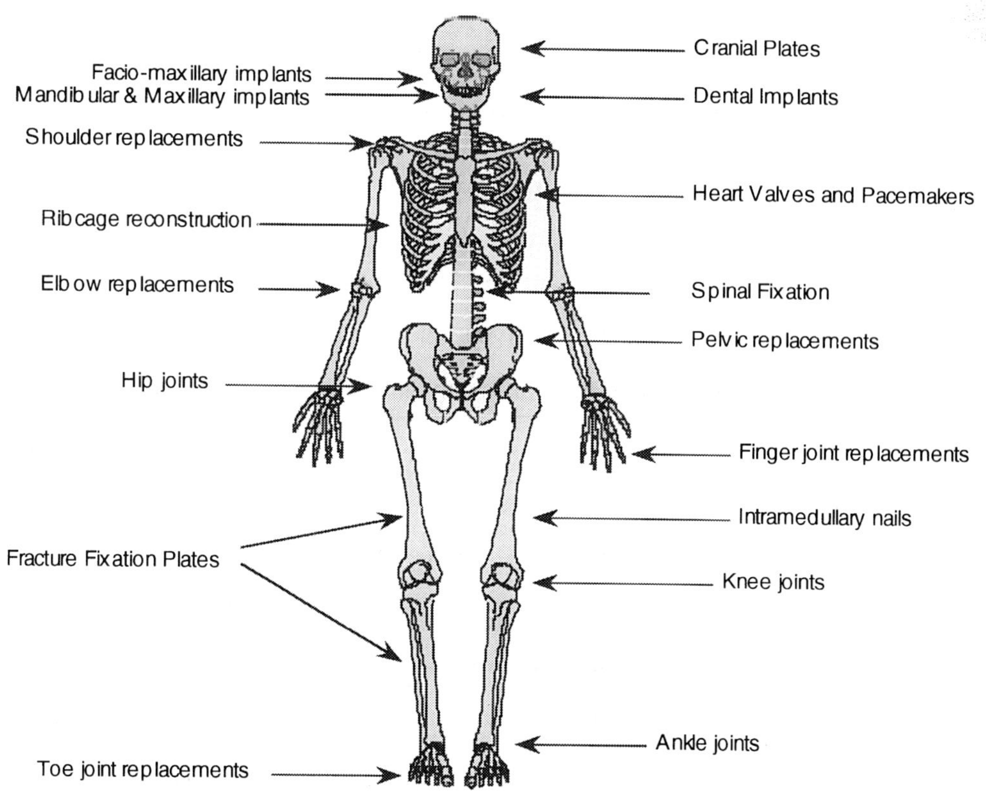

Figure 1. Titanium alloy implants have been used extensively throughout the body

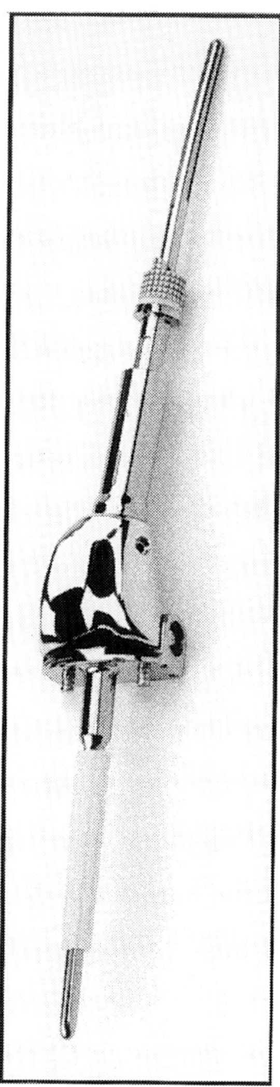

Plate 1. An extendible distal femoral and knee replacement. The implant shaft and femoral intra-medullary stem are fabricated from titanium alloy (Ti-6Al-4V) whilst the knee and tibial components have been cast in cobalt chrome molybdenum. The grooved structure at the implant shoulder adjacent to the femoral intra-medullary stem has been coated with hydroxyapatite ceramic to encourage bony in-growth to aid fixation.

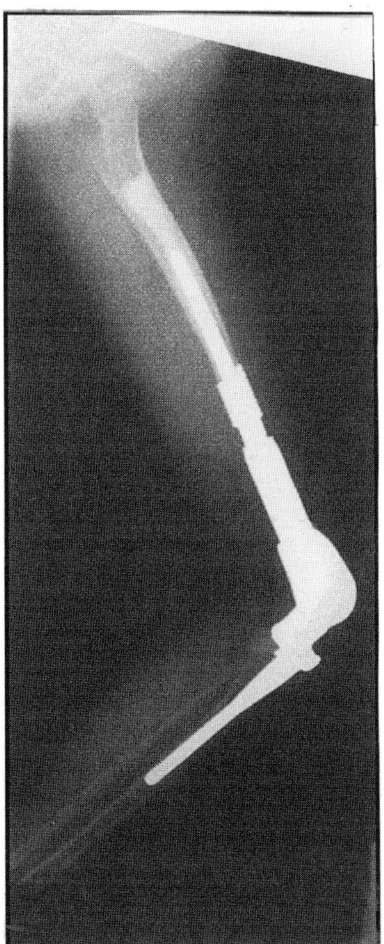

Plate 2. A radiograph (lateral view) of a young girl showing an extendible distal femoral and knee replacement. The implant has secured into the femoral medullary canal using a cemented intra-medullary stem. Over 350 extendible replacements have been used for juvenile patients who have had bone tumour.

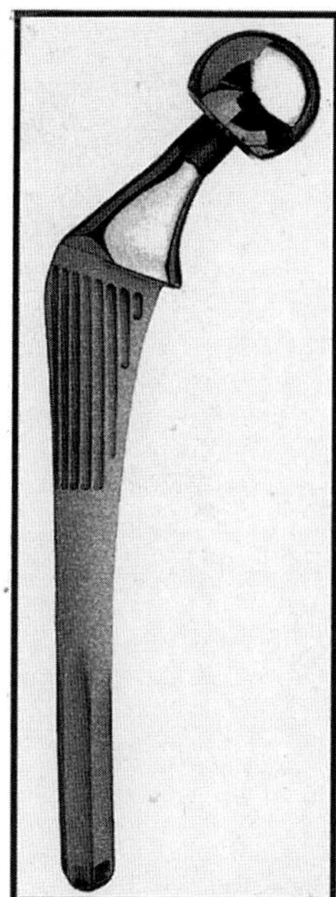

Plate 3. A custom-made CAD-CAM hip replacement. The femoral hip stem body is fabricated from titanium alloy (Ti-6Al-4V) and is coated with a thin layer (80 μm) of hydroxyapatite ceramic. The cobalt chrome molybdenum femoral head is modular. Over 1,300 CAD-CAM hips have been inserted over the last decade most of which have been used to treat failed standard joint replacements or complex primary bone conditions.

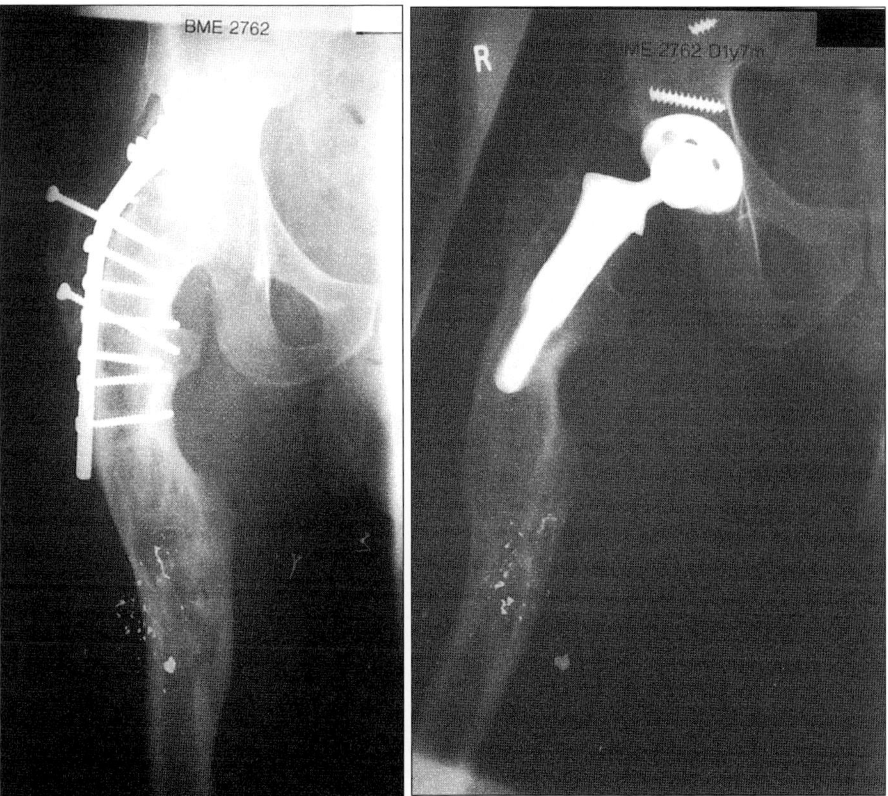

Plate 4. The radiograph on the left shows a complex reconstruction problem. The patient's femur is malformed and previously the hip joint had been fused. Subsequently the fusion was reinforced using plates and screws. Using CAD-CAM technology a custom-made hip replacement was designed and manufactured to fit the abnormal proximal femoral canal. The patient regained hip motion following the surgery.

S696/003/99

Development in high-productivity welding of titanium

P L THREADGILL, M F GITTOS, and **L S SMITH**
The Welding Institute, Cambridge, UK

INTRODUCTION

The market for titanium is increasing steadily as its properties of low density, high strength and good corrosion resistance become increasingly exploited. The uptake of any engineering alloy is influenced by the ease with which it can be joined. Although the common titanium alloys such as commercially pure titanium and Ti-6Al-4V are not difficult to weld by fusion processes, the methods in common use are generally very slow, particularly if joints of the highest integrity are required. The most common welding process, TIG welding, is also prone to environmental conditions, and is best operated in an enclosed welding shop where the ambient temperature and contaminant levels can be closely controlled. The need for closely controlled gas shielding, and the intolerance of titanium to contamination by other impurities (e.g. iron) has given it a reputation for being difficult to weld. Titanium is one of the easiest metals to weld by TIG, providing that appropriate precautions are taken. Although TIG welding has been the favoured method for joining titanium, the process is slow, and there have been many demands for higher productivity processes in which quality is not compromised. Indeed, the industry has also been demanding alternative high quality joining processes for use where TIG is inappropriate, and where productivity is not a burning issue.

In almost all of the areas where titanium alloys are used, there is a large potential cost saving available by introducing higher productivity methods which can still guarantee welds of a very high quality. The purpose of this article is to review some of the more productive methods now available, and comment on their relative strengths and weaknesses.

The processes described will not include all available processes. Further information is available in numerous other sources (1,2,3,4,5).

SOLID STATE PROCESSES

There are a number of solid state processes used for joining titanium. These include friction welding, flash butt welding, diffusion bonding, explosive welding and MIAB welding. With some of these, for example friction welding, there are many process variants. Some of the most interesting are described below:

Rotary friction welding: This process is far from new, but it is seldom applied to titanium alloys. In this process, a component with rotational symmetry is rotated under pressure against a static component. Frictional heating will soon cause both components to soften, and the softened area will be expelled as flash. When the interface is at a sufficiently high temperature, the rotation is rapidly stopped, and a higher forge pressure applied to complete the joint. In this process, as in all solid state processes, there is no fusion of the material. Since the process is fully mechanised, there is no dependence on manual skills in making the weld, although care must of course be exercised in setting up the equipment. Advantages of this process include short welding time, typically a few seconds even in large sections, no filler wire, and no shielding gas. There are two principal variants, namely continuous drive friction weld, and inertia friction welding. In the former, one component is attached to a motor which is continuously driven, and when the resultant heating and axial loading has caused a pre-determined burn-off off material, the rotation is stopped rapidly, and a higher forge force is usually (although not always) applied. In inertia friction welding, the rotating component is attached to a flywheel, and the joint acts as a brake when the two components are brought together. It can be seen that in continuous drive variant, the rate of energy input into the joint is constant, whereas in inertia welding, the rate decreases as the rotation speed decreases. Figure 1 shows the appearance of a section through a continuous drive rotary friction weld in a 14mm thick Ti-6Al-4V-Pd pipe of 246mm diameter.

Linear Friction Welding: In this process, the relative motion between the two components is a linear reciprocating motion, typically at rotation speeds of 25-100Hz, and with an amplitude of ±3mm. The welding time is typically 2-3 seconds. So far, the process has only been used in the aircraft engine industry for joining blades to disks, but recent developments in machine design, which look likely to lead to a reduction in machine costs, have re-awakened interest in many other sectors, particularly the automotive industry. The properties are good, with the mechanical strength of the weld being similar to that of the parent material in most cases.

Radial Friction Welding: Radial friction welding overcomes the requirement to rotate long lengths of pipe by using a consumable ring and completing two welds simultaneously (one between the consumable and each of the stationary pipes). The consumable ring is rotated, and simultaneously compressed radially to give a weld.

The main advantage of radial friction welding over a rotary friction welding process is the removal of the need to rotate long lengths of pipe. Internal flash is also eliminated, although a small change in internal profile, where the pipes are compressed against the mandrel, will typically be present. Its disadvantages over rotary friction welding include the increased complexity and, therefore, cost of the welding machine and the added difficulties in monitoring and controlling weld quality, since two welds are produced rather than one. Even so, the process is fully mechanised and quality control will probably be easier than for fusion welding processes.

Homopolar Welding: This technique has received considerable publicity, but has not been used in commercial production. Homopolar welding is a new method currently under development in the USA, where it has been developed primarily for welding steel pipes. In this process, kinetic energy stored in a flywheel is rapidly converted to a high direct current low voltage electrical pulse using a homopolar generator, and this high current pulse is passed across a closely butted weld joint, causing a resistance weld to be made. A high axial load is also applied, causing softened material to be expelled. In essence, the homopolar system acts as a huge capacitor. Like radial friction welding, neither of the components has to be rotated, and as with all the friction processes described above, no shielding gas is required, even for titanium. Although most work to date has been on steel, preliminary trials have been undertaken on titanium, apparently successfully, although no published data are available.

FUSION WELDING

Although fusion welding, in the form of TIG welding, is the most widely used process for joining titanium, there are two issues which need to be addressed, namely porosity and contamination.

Porosity in titanium fusion welds can be formed for a variety of reasons, but the most insidious is undoubtedly the profound influence of the condition of the joint surfaces. A more detailed description of weld metal porosity and its prevention in titanium fabrications is given elsewhere (1,6) but the evolution, coalescence and entrapment of hydrogen bubbles during solidification is probably the most common cause. The joint surfaces should be scrupulously clean prior to welding, since even a fingerprint can lead to porosity. However, cleanliness alone is insufficient. The rutile (TiO_2) scale that imparts such considerable corrosion resistance is hygroscopic and adsorbs moisture from the environment. The hydrated layer provides a source of hydrogen and steps should be taken to minimise its presence. This includes using procedures that reduce surface roughness and the hydration state (i.e. x in $TiO_2.xH_2O$) prior to welding. Machining without aqueous lubricants and welding in the same 24h period are consequently useful strategies.

One interesting facet of porosity in titanium weld metal is its seeming dependence on the presence of a prior joint surface. Autogenous melt runs on the surface of titanium plates are usually free of porosity, seemingly independent of the presence of grease, oil and other contaminants. Likely reasons for this are discussed elsewhere (6), but it is tempting to speculate that this behaviour could be exploited to good effect in eliminating near-surface pores in fusion welds. Much development work is necessary but, potentially, many of the concerns regarding porosity could be countered simply by applying autogenous melting passes over the completed weld surface. Indeed it has been demonstrated that remelting by TIG and EB welding can decrease porosity in a previously deposited bead (7).

Porosity can also adversely affect fatigue properties. It is therefore most desirable to eliminate it, but it is also clearly important to be able to detect it. The dramatic effects of porosity on fatigue life have been discussed by Lindh and Peshak (8) who showed that porosity can have a marked influence on the fatigue properties of titanium alloys, particularly when near the surface. They demonstrated that, in the absence of defects and other stress raisers, fatigue cracks invariably initiated form pores. This was true regardless of pore size, although smaller pore sizes were less detrimental than larger ones.

Titanium is an extremely reactive material at high temperatures and in the liquid state. In particular rapid reaction will occur with the interstitial elements oxygen, nitrogen, carbon and hydrogen. Thus, the need for scrupulous cleanliness, together with an effective gas shield is apparent. The levels of atmospheric contamination are normally related to oxide colours, but care is required in the interpretation of the colours. A perfectly silver appearance does not necessarily guarantee freedom from contamination, as such a surface can be formed if a good trailing shield is used after a poor arc shield. Similarly, a highly coloured surface may indicate only a poor trailing shield. The depth of contamination is probably only slight if contamination occurred only in the solid state. This problem of detection of contamination is difficult, and is the subject of considerable research efforts. The currently accepted significance of surface colour of titanium welds is given below.

Table 1. A rule of thumb classification system for judging the success of gas shielding of titanium welds

Colour of weld zone	Interpretation
Silver	Correct shielding, satisfactory
Light straw	Slight contamination, but acceptable
Dark straw	Slight contamination, but acceptable
Dark blue	Heavier contamination, may be suitable depending on service
Light blue	Heavy contamination, unlikely to be suitable
Grey blue	Very heavy contamination, unacceptable
Grey	Very heavy contamination, unacceptable
White (loose deposit)	Very heavy contamination, unacceptable

Guidelines on the various processes available, together with their advantages and disadvantages, have recently been reviewed and collated into a handbook on titanium welding, published by the Titanium Information Group and TWI. Readers are referred to this for more detailed information. However, the following summary outlines some of the more recent developments in welding of titanium. It is not intended to be a complete catalogue of the available processes. An example of a multipass TIG weld in Ti-6Al-4V is shown in Figure 2.

Keyhole Plasma Welding: Strictly speaking, this is not a new process, but it remained little known for many years. The advantage of the process is that it can be used to make single pass welds in quite thick material, up to 15mm in a single pass. The keyhole plasma welding process offers great potential for joining titanium risers and flowlines, its main advantages being productivity and an apparently decreased susceptibility to porosity when compared with TIG welding. The process is broadly similar to TIG welding, but a copper nozzle, through which flows a portion of the shielding gas (commonly referred to as the plasma gas), constricts the arc. A portion of the plasma gas is ionised forming a fully penetrating plasma jet. Single pass welds up to 20mm thick are feasible in the flat (1G) position, allowing dramatic productivity gains to be achieved (9,10,11). Trials performed at TWI have shown that joint surface conditions, known to give porosity in TIG welds, show no such tendencies for keyhole plasma welds. Other advantages over TIG welding include a reduced weld zone grain size and the simple square butt joint geometry required by the process.

A typical example of a keyhole plasma weld in 15mm thick Ti-6Al-4V is shown in Figure 3.

Reduced Pressure Electron Beam (RPEB) Welding: Electron beam welding is firmly established in the aerospace and other high quality sectors for which joint performance is critical. Narrow welds can be produced reproducibly with little distortion and very good control can be maintained on weld geometry. Whilst conventional EB welding seems inappropriate for use offshore and for many other requirements, reduced pressure EB (RPEB) welding processes (12) have been developed which are suitable for welding large pipelines in-situ. Conventional EB welding is performed under a medium to high vacuum, whereas RPEB is carried out in a partial pressure of helium. Thus RPEB can be performed in a chamber environment which seals to the work piece, rather than the fully enclosed chamber required for conventional EB welding. The process does not require a high vacuum capability and so mechanical vacuum pumps are more than adequate. Equipment for laying steel pipes using RPEB is under development, but satisfactory RPEB welds have been made in titanium pipe in material up to 16mm thick (11). A typical example is shown in Figure 4.

The advantages of the RPEB process are similar to those of keyhole plasma welding, although many of the difficulties, such as weld closure, have been comprehensively solved for EB welding. A possible disadvantage lies in the rapid cooling rates experienced which can produce martensitic microstructures. Whilst the martensite formed in Ti-6Al-4V is not as hard as that formed in more heavily β-stabilised titanium alloys, fracture toughness can be lowered compared with the base material (13). However, preliminary toughness investigations suggest that RPEB weld metal performance is similar to that of the base material (11). Electron beam welds are susceptible to porosity and all precautions necessary for its avoidance in TIG welds apply equally to RPEB welding.

Laser Welding: Laser welding has been successfully applied to titanium alloys (e.g. 14). Penetration is particularly good for titanium, due its low thermal conductivity and high absorption rate for infrared light. The process offers similar advantages to EB, but the process operates at atmospheric pressure, albeit with local shielding. Laser welding will also produce similar microstructures to RPEB, i.e. alpha prime martensite within prior β grains. Laser welds are susceptible to porosity, but can also be prone to the entrapment of the helium shielding gas. Adequate joint preparation strategies will minimise the former, whilst careful selection of welding parameters should eliminate the latter.

DISCUSSION

As stated in the Introduction, titanium is generally not a difficult material to weld providing that due care is paid to its very high reactivity and the ease with which it can become contaminated. The precautions are more stringent than those required for most other materials, and this has given titanium a reputation for being difficult to weld. This is undeserved, although it does require great attention to fine detail, in particular with fusion welds. With most forge welding processes, the requirement for additional precautions is minimal. It is also apparent that it is possible to make welds of very high quality by a variety of solid state and fusion processes, and it is not necessary to sacrifice quality in order to achieve high productivity. The Table below summarises some of the more common processes, but is not intended to include all possible processes. Those included are typical processes in production, together with other possibilities which are in lower volume use, or which have been the subject of considerable recent development.

Table 2. Attributes of the Welding Processes

Process	Position	Shielding Gas	Development Effort Needed*	Weld Time	Capital Costs	Susceptibility to Weld Porosity
TIG	Any	Yes	None	Slow	Low	High
MIG	1G, 2G	Yes	Low	Slow	Low	High
Keyhole Plasma	1G, 2G	Yes	Moderate	Medium	Low	Low
RPEB	2G	partial vacuum	High	Medium	High	High
Laser	1G, 2G	Yes	Low	Medium	High	High
Rotary Friction	Any	No	Low	Fast	High	Immune
Radial Friction	Any	No	Moderate	Fast	High	Immune
Linear friction	Any	No	Low	Fast	High	Immune
Diffusion Bonding	Any	vacuum	Low	Slow	Medium	Immune
Homopolar	Any	No	High	Fast	High	Immune

Notes: * The development effort required is of course dependent on the final application. This Table is therefore rather subjective.

In the above Table, the weld time refers to the time during which the welding process is operational. It does not include set-up and knock-down times, and these must of course be included in any assessment of the total productivity of the process. As this will be component specific, no attempt has been made to include relevant data here. Capital costs are again subjective, but fully mechanised processes allow high productivity, and the costs can be justified, in particular for high added value products, which represents the majority of the titanium fabrication market.

Although TIG is not very productive, it scores over all the other processes in flexibility. It can cope with almost any size and shape of components, whereas the more productive processes generally do not have this capability.

An example of comparative productivity has been given by Smith et al (4) for welding titanium pipe. This data is given in Table 3 below.

It is evident that use of processes other than TIG or MIG can lead to better productivity, and also to elimination of porosity in many cases. However, these advantages have to be set against much higher capital costs, and the relative lack of maturity of these processes for welding titanium.

Table 3. Approximate joint completion times for 250mm outside diameter, 14mm thick titanium pipe, including setting up.

Process	Joint Completion Time (hours)
TIG	10
MIG	6
Keyhole Plasma	1
RPEB	1
Laser	1
Rotary Friction	1
Radial Friction	1

SUMMARY

The key issues in welding titanium and its alloys are primarily the avoidance of porosity and contamination, and in addition the achievement of improved productivity. It is hoped that this paper has demonstrated, at least in outline, that welding titanium should hold no fears for the experienced fabricator, and the opportunities for exploiting several more productive processes which also avoid porosity and contamination are there for the taking. The data presented indicate that significant improvements in titanium weld quality and productivity should be achievable.

ACKNOWLEDGEMENTS

Some of the work referred to in this document was funded by Industrial Members of TWI as part of TWI's Core Research Programme.

REFERENCES

1. Smith L S, Threadgill P L, Gittos M F and Peacock D A; "Welding of Titanium". Titanium Information Group Brochure No. 6. April 1999.

2. Threadgill P L; "The Potential For Solid State Welding Of Titanium In Offshore Industries." Proc Seminar on Right Use of Titanium III, Stavanger, Norway, 1997.

3. Gittos M F, Threadgill P L and Smith L S; "Joining of Titanium Risers". Proc 17th Int. Conf. On Offshore Mechanics and Arctic Engineering. Lisbon, July 1998, OMAE98-2213.

4. Smith L S, Gittos M F and Threadgill P L; "High Quality and Productivity Joining Processes and Procedures for Titanium Risers and Flowlines". Proc Conf Titanium Risers and Flowlines, SINTEF, Trondheim, Norway, February 1999-11-06.

5. Webster R T; "Welding of Titanium Alloys". ASM Metals Handbook, 10th Edition, v6, 783–786, 1993.

6. Smith, L S and Gittos, M.F. "A Review of Porosity and Hydride Cracking in Titanium and its Alloys." TWI Members' Report 658/1998.

7. Gorschkov A I: "Aspects of Pore Formation in the welding of Titanium". Svar. Proiz., no. 7, 21–23, 1968.

8. Lindh D V and Peshak G M; "The Influence of Weld Defects on Performance". Welding Journal **48**, (2), 45s–46s, 1969.

9. Boucher C, Messager F, Heuze J-L and Gaillard F; "Open-air Arc Welding of Medium Thickness Titanium Alloy". Proc Eurojoin 2, Second Conference on Joining Technology, Florence Italy, May 1994 (Italian Institute of Welding).

10. Ramsland A R, Ødegård J, and Hauge M; "Welding and Mechanical properties of High Strength Titanium Alloys for Offshore Applications". Proc Int Conf on Joining and Welding for the Oil and Gas Industry, Oct 1997, TWI.

11. Smith L S and Gittos M F; "High Productivity Pipe Welding of Ti-6Al 4V alloys. TWI Members Report 660/1998.

12. Belloni A and Punshon C S; "Reduced Pressure EB Welding for Offshore Pipelines". IIW Document IV-680-1997.

13. Boyer R, Welsch G and Collings E W; "in Materials properties handbook; Titanium Alloys, ASM International, 483, 1994.

14. Dabiezes B, Menager B and Diani J M; "Laser Welding TA6V Titanium Alloy of Thickness 20mm". CISSFEL Int. Conf. On Welding and Melting by Electron and Laser Beams., Publ. Commissariat a l'Energie Atomique, la Baule, France, June 1993, **1**, 251–258.

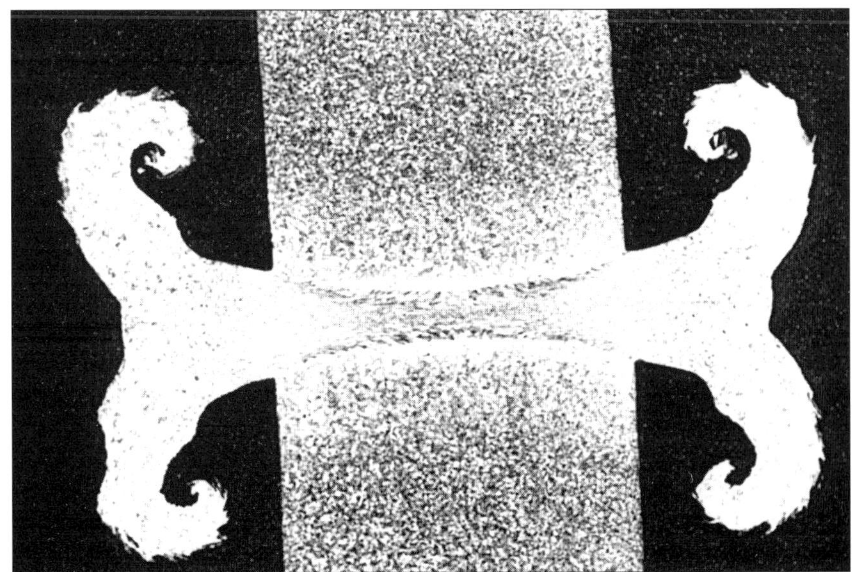

Figure 1. Section through a continuous drive rotary friction weld in a 14mm thick Ti-6Al-4V-Pd pipe, x4.5. (Neg AK1208)

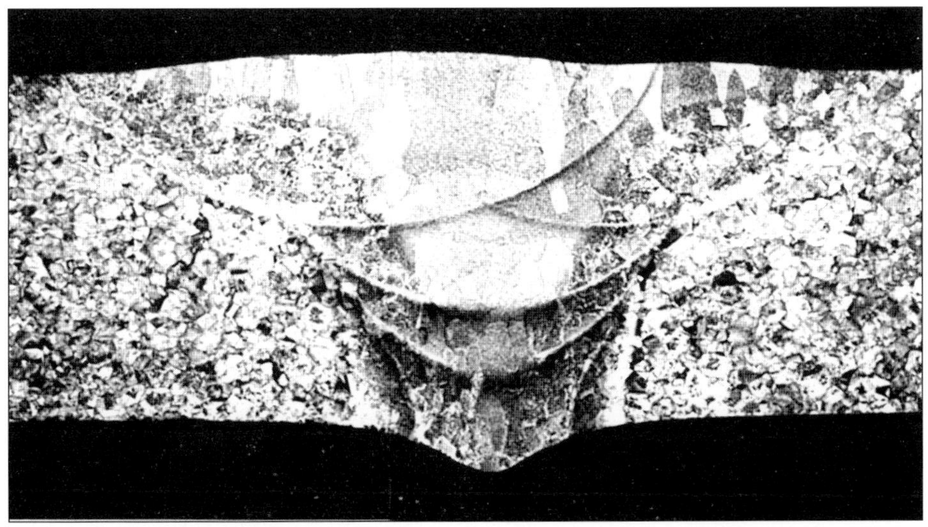

Figure 2. Multipass TIG weld in Ti-6Al-4V. x5.(negs AK1724, 1725)

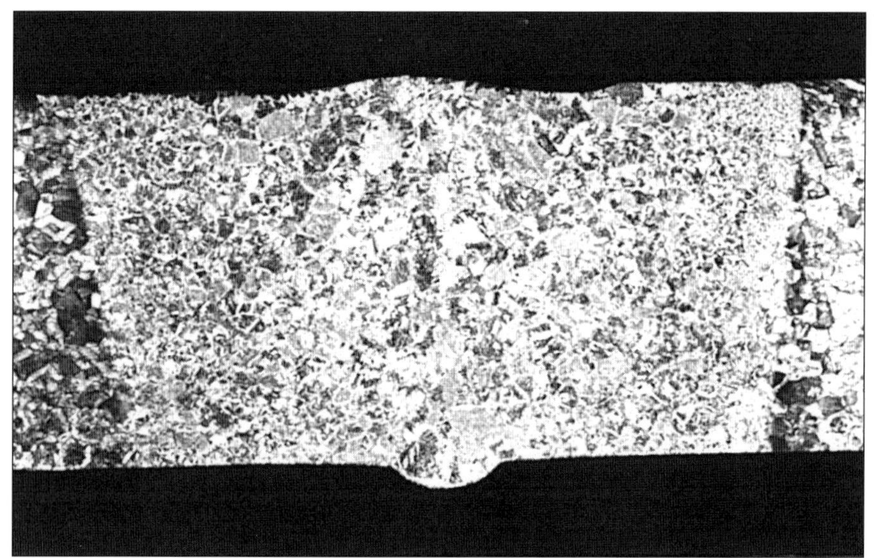

Figure 3. Section through a keyhole plasma weld in 15mm thick Ti-6Al-4V. x5, (Neg AK611)

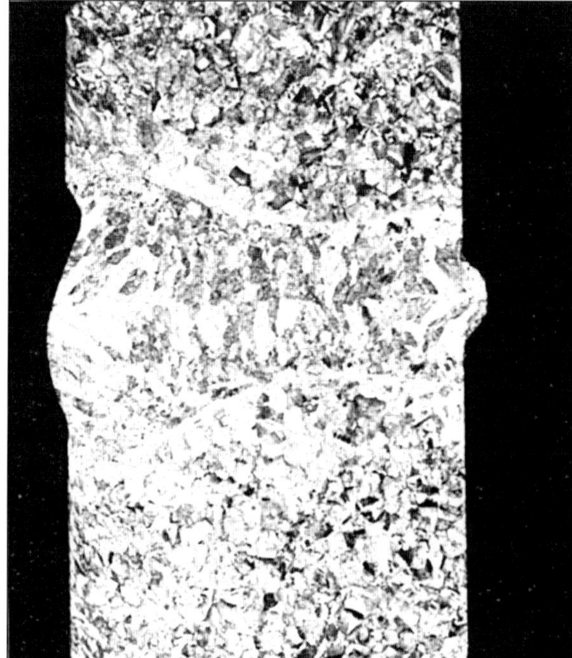

Figure 4. Section through RPEB weld in 15mm Ti-6Al-4V, x5. (Neg AK1722)

S696/004/99

Manufacture of titanium alloy components for aerospace applications

C Q HUGHES, P J BRIDGES, and **P S BATE**
DONCASTERS plc, Melbourne, UK

INTRODUCTION

The advantages of titanium alloys in aerospace are well known. They offer high strengths and good corrosion resistance, together with useful performance at temperatures up to about 600ºC with a density about 60% of engineering steels and almost half that of nickel alloys. Although there are problems with its cost and supply – and the perception of it as an 'exotic' metal – titanium finds extensive use in aerospace. Certain aspects of titanium mean that the effective manufacture of components requires special techniques, or at least a deal of care. DONCASTERS plc has a long history in the forming of titanium alloys by forging, sheet forming, casting and machining. About 250 tonnes per annum of such alloys are processed across the Group to make aerospace components.

Pure titanium melts at 1675ºC, somewhat higher than pure iron. In the solid state it exists in two allotropic forms. The low temperature one, known as α titanium, has a close-packed hexagonal crystal structure. Above 880ºC in the pure metal, this changes to a phase with a body-centred cubic structure known as β titanium. Alloying elements tend to show a chemical preference for one or other of these phases. For example, aluminium has much greater solubility in α titanium and tends to stabilise that phase, whereas vanadium stabilises the β phase. Adding both α-stabilising and β-stabilising elements leads to alloys in which the two phases co-exist over a wide temperature range- though in varying proportions. The composition Ti- 6%Al- 4%V gives the commonest titanium alloy. Most of the other titanium alloys important in aerospace are essentially of the same type, although many have a greater level of α stabilising elements and some are heavily β stabilised.

Cold forming of most alloys is not feasible, and hot working is required. Hot working also allows the generation of microstructures which cannot be produced by heat treatment alone. It is difficult to restrict grain growth at high temperatures where alloys comprise entirely of β phase, and cooling – at moderate rates – from temperatures leads to the precipitation of α phase in the form of plates. Hot working at temperatures where both phases are present 'breaks up' the α plates and leads to a fine and uniform microstructure. Such a structure is generally preferred, either as-worked or as a basis for further heat treatment.

In the context of manufacturing, perhaps the most important feature of titanium and its alloys is its chemical reactivity. In particular, titanium combines with oxygen and nitrogen to form thermodynamically stable compounds. At relatively low temperatures, say below 600°C, these compounds occur as thin, self-healing surface films which lead to exceptional corrosion resistance. However, at higher temperatures titanium has the unfortunate property of dissolving – 'eating' its own oxides or nitrides. In this way it is different to, say, aluminium, where the protective oxide surface exists even in the liquid state. A piece of titanium alloy exposed to air at 400°C will remain effectively inert indefinitely, but exposure at 1200°C will result in a friable mass of oxide and nitride within a few hours. The fact that the reactions are exothermic is a further problem, and certainly could cause disaster if liquid titanium were exposed to air. The problem should not be overstated- it is possible to flame cut titanium for example, and argon shielded welding is straightforward- but it cannot be ignored. Casting, forging and most sheet forming operations on titanium alloys require high temperatures and clearly the metal has to be protected from normal atmospheric elements in some way.

Although this reactivity causes problems in hot processing, it also creates an opportunity. As will be discussed below, the clean surface produced by dissolution of surface oxide and nitride layers allows diffusion bonding which, in combination with superplastic forming, means that complex sheet structures can be produced more easily in titanium alloys than in other engineering metals.

The reactivity of titanium at elevated temperatures is not restricted to atmospheric elements. It will readily dissolve, or at least bond with, most other metals and this means that machining operations have to be done in a way which is somewhat different to those for engineering steels.

The value of titanium swarf, off-cuts and grindings is considerably less than that of the input material and these residues cannot be easily re-cycled without expensive recovery treatments. Therefore there is considerable economic incentive in the primary manufacturing stage to minimise material input and make as near as possible to the required form. This is graphically illustrated in the evolution of methods used to make titanium blades for gas turbines. These were originally machined from solid bar, with the associated waste of material. There was then a move to using oversize forgings from which the final shape is machined. More recent developments in preform design using mathematical modelling, lubrication technology, tool heating and process modelling and control have enabled precision airfoils which need very little treatment after forging

FORGING AND RING ROLLING

The majority of airfoils in the compressor stages of gas turbines for aeroengines are made from titanium alloys. As noted above, it is desirable to make these components near to net shape, i.e. precision forged (Figure 1). Prior to that stage, the cast ingot material is forged to produce bar with not only the required geometric form but also to modify the microstructure. A certain amount of deformation is required to give the structural breakdown described above, and final shape forming operation may not achieve this in regions such as the root block of an airfoil.

To minimise reaction with normal atmosphere, a glass coating is used. This is only necessary in the later stages when section sizes become small. Material affected by pick-up of atmospheric elements – the so called 'alpha case' needs to be removed by mechanical or chemical means. The glass coating also acts as a lubricant.

Conventional forging operations involve a large initial temperature difference between the dies – usually tool steel – and the workpiece. The dies are typically at about 200ºC with the workpiece being pre-heated to between 900ºC to 1000ºC. The forging operation needs to be quick, otherwise the workpiece cools too much. That would not only increase forging load and possibly lead to damage to the dies, but could also lead to cracking or other damage to the workpiece. In terms of size range covered we make gas turbine blades from 2 to 75 cm long with chord widths from 1 to 25 cm. These would be done on screw presses with capacity up to 15,000 tonnef and could be either precision forgings or oversize forgings. The latter would go for final shaping by machining or electrochemical machining. The larger titanium blades would be for the fan at the front of the engine whilst the smaller ones would generally be for the compressor stages.

Hot forging is also used to make titanium rings and engine casings. The method used in manufacture of these parts depends on their geometry. Large section rings and casings would be formed from a pierced or punched billet on a press using a series of support tools. Smaller sections would be ring rolled from a preform and the end product could be either rectilinear in section or have some shape; examples are shown in Figure 2. Using a combination of these methods it is possible to make titanium alloy rings out to a diameter of 2.5 m, a section size of 125 mm, and a weight of 1200 kg. These components have the advantage of containing the optimum mechanical properties because they have been forged. Their disadvantage is that there is often a lot of machining needed to reach the final required shape.

Under certain circumstances, it is necessary to form material at a slower rate than occurs in conventional forging operations. To do this, the temperature difference between dies and workpiece needs to be reduced by heating the dies to the same (isothermal forging) or a somewhat lower (hot die forging) temperature. One case where isothermal forging has been used is in the manufacture of poppet valve blanks in titanium aluminide intermetallic alloy, where the very limited ductility of the material at temperatures below about 700ºC means that chilling has be avoided. In that case molybdenum alloy tooling and a vacuum were used. Airframe parts in high strength titanium alloys can be forged isothermally using nickel alloy dies, using the 3200 tonnef isothermal press at DONCASTERS Sheffield Precision forge. There is greater microstructural control with isothermal forging than with conventional operations as well as lower tool and press loads, though the operation is, of course, slower.

SHEET FORMING

As with other forming operations, sheet forming often needs to be carried out at elevated temperatures. Some β, and very lean, alloys can be formed cold, but Ti-6%Al-4%V and others require hot forming. Clearly the geometry means that such processes have to be essentially isothermal, and a large proportion of sheet forming with titanium is carried out at temperatures from 800ºC to 950ºC using gas pressure to effect the operation. At these temperatures, the flow stress of the alloys is low and so the pressures are moderate. The low flow stress also means that residual stresses and springback are low, and this type of process gives good shape definition. Hot gas-pressure forming of large aircraft exhaust components in titanium alloys is carried out at DONCASTERS Bramah, with excellent dimensional control.

With many alloys, slow forming at temperatures of about 900ºC-950ºC gives extremely high ductility – 'superplasticity'. In many cases forming can be carried out quicker and at lower temperatures, but superplastic forming (SPF) allows very complex shapes to be formed. This is particularly true when it is combined with diffusion bonding (DB).

The ability of titanium to dissolve surface oxide and nitride layers at elevated temperatures means that clean metal surfaces can be produced simply by heating in inert atmospheres. At these temperatures the plasticity of material is high and quite modest loads can give intimate contact between two alloy surfaces, which then bond. These diffusion bonds are essentially perfect: they cannot be microscopically detected. The fact that bonding can be prevented through the use of 'stop-off' mixtures of ytrria and hexagonal boron nitride allows the manufacture of multilayer forms which can then be inflated in hot gas-pressure forming to form complex shapes. It is also possible to allow surfaces to come into contact during SPF which then bond.

MACHINING

Despite its reactivity and tendency to bond, titanium alloys can be machined by all the standard processes used for engineering steels. However, it is imperative to keep the temperature at the tool tip low during cutting operations, and this is not helped by the relatively low thermal conductivity of titanium alloys. Cutting speeds need to relatively low, and a suitable halogen-free lubricant used. Carbide tools are preferred and ceramic coated carbides are even better.

Grinding can be carried out, again with adequate lubrication and relatively low speeds. Both silicon carbide and alumina wheels are suitable. It is particularly important in grinding to prevent fire in grinding fines, and precautions should also be taken with machining swarf.

Alloys can be electro-discharge machined and chemically- or electrochemically- milled. The latter process is useful for generating high- precision forms on airfoils. Chemical milling, using solutions of hydrofluoric and nitric acids is used for removing alpha case, and, in conjunction with masking compounds, for selectively reducing the thickness of sheet.

CASTING

The problems caused by the reactivity of titanium become most acute in casting. Not only does the process needs to be carried out in vacuum, or an inert atmosphere, it also needs special melting facilities. Very few refractories are suitable for melting titanium, and those are either extremely expensive or very difficult to deal with in practice. The usual solution is to effectively use a titanium crucible in a method known as skull melting. At DONCASTERS titanium foundry, SETTAS SA, deep pool electric arc melting is used, as shown in Figure 3.

The relatively low thermal conductivity of the solid titanium means that a large liquid mass can be produced with quite a thin solid 'skull', maintained by heat conduction into the water-cooled copper crucible.

This melting technique means that the liquid metal temperature cannot be much greater than the melting point i.e. the superheat possible is very low. Because of this, it is imperative to fill moulds as quickly as possible. The technique adopted at SETTAS is centrifugal casting. This employs casting tables of diameter up to 3 m rotated to give peripheral speeds sufficient to generate accelerations of about 50 g. The furnace charge can be up to 1 tonne of metal. Moulds- either sand, ceramic shell (investment casting) or metal- are usually filled from runners returning inwards which are supplied with metal *via* 'legs' radiating from a central sprue. Casting is rapid- a full charge would be poured in about 5 s. The rapidity of casting minimises time for metal- mould reactions. The centrifugal system helps to consolidate the castings while they are solidifying and thereby enhance their integrity.

The centrifugal casting arrangement is ideal for making near-axisymmetric shapes such as engine casings. An example is shown if fig. Y, which was cast into a ceramic shell mould. This type of component is conventionally fabricated from rings and extensively machined, and a great cost saving results from using casting.

Casting defects such as shrinkage porosity can be eliminated by hot isotatic pressing (HIPing), and no included oxide contamination occurs in titanium casting because the molten metal is contained by solid metal. However, the thermomechanical refinement of microstructure is not easily done in a casting and there is a trade-off in terms of reduced mechanical properties, particularly toughness and fatigue resistance.

Within DONCASTERS there is a unique capacity for casting, forging and fabricating titanium alloys, and this gives opportunities for future production of cost-effective aerostructural assemblies as an alternative to one-piece castings.

SUMMARY

The paper has reviewed the range of capabilities within the DONCASTERS Group for making titanium components for aerospace applications. The methods can be complementary or competitive. The design of a component has a considerable bearing on the choice of manufacturing route and associated costs. To make the most of the opportunities for titanium it is essential that design, engineering and manufacturing are considered together at the concept stage.

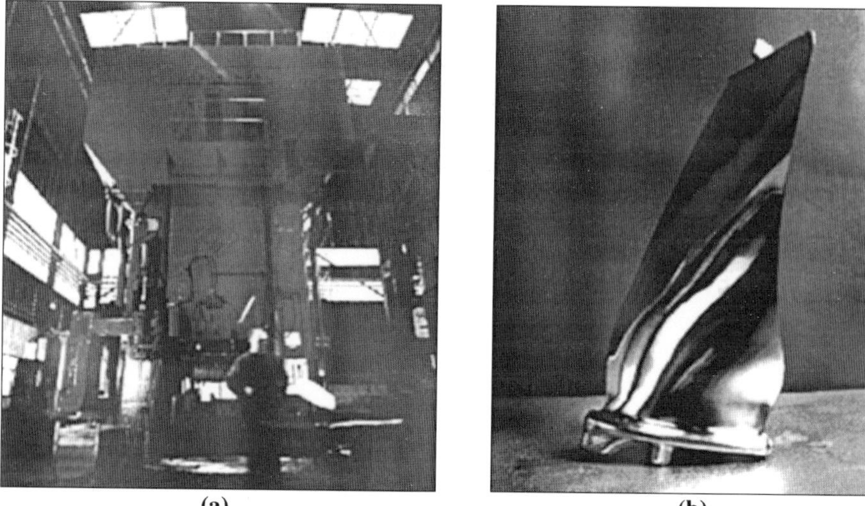

(a) (b)

Figure 1. (a) An airfoil being removed from the forging press and (b) a large compressor blade made by precision forging.

Figure 2. A selection of ring sections made by forging, ring rolling and machining.

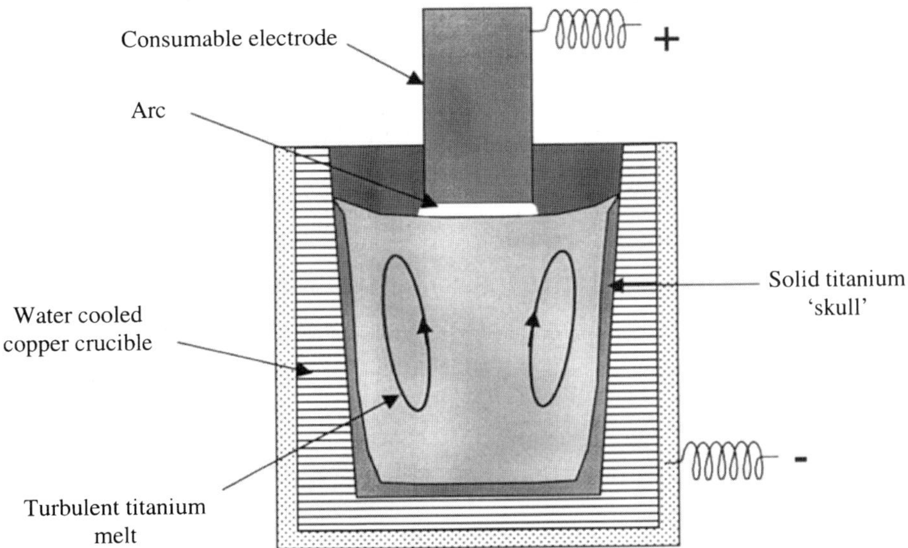

Figure 3. **A schematic of the vacuum arc melting arrangement used for titanium casting. The arc is pulsed DC, this causes vigorous stirring and gives very good homogeneity.**

Figure 4. Engine casing centrifugally cast into shell ceramic mould.

S696/005/99

Titanium drilling riser technology

P JAQUES
Oil States Industries (UK) Limited (formerly Hunting Oilfield Services Limited), Aberdeen, UK

ABSTRACT

The use of titanium and titanium alloys continues to increase in a wide range of marine and offshore applications. Oil States Industries was the EPC contractor for the initial prototype titanium drilling riser and the world's first titanium alloy Ti-6Al-4V ELI (Grade 23) high pressure drilling riser. During the prototype phase a comprehensive material test programme was carried out to validate the material selection. Assessment of stress corrosion cracking in air and sea water, evaluating wear resistance, evaluation of manufacturing and service induced defects on fatigue and crack growth behavior when subjected to the operating environment. This paper presents an overview of the design, manufacture and test methods used for the titanium drilling riser to meet the functional and operational design requirements. The paper concludes with a discussion on a recent development of a cost effect titanium and composite drilling riser for use on Heidrun.

INTRODUCTION

Titanium alloys have a 45 year history of use in the aerospace and defence industries. In the past decade there has been a substantial growth in the use of titanium alloys offshore, applications for heat exchangers, fire and ballast water piping, production tubing and casing, production riser stress joints and well intervention risers. A taper stress joint was supplied to the Gulf of Mexico for Placid Oil in the Green Canyon field in 1987. The joint was retrieved in 1989, no damage, refurbished, stored and then installed offshore for Ensearch, Garden Bank in July 1995. A total of fifteen Ti-6Al-4V titanium taper stress joints have been subsequently delivered for the Oryx Neptune field. The first extensive use of titanium has been for the Conoco high pressure drilling riser on Heidrun Tension Leg Platform. Project specific design and operational constraints inherent to the Heidrun TLP motivated the use of titanium for this application over the traditional alternative of steel. In particular requirements relating to riser interference in severe weather conditions given the wellbay spacing

constraints on the production template. The selection of titanium provided a substantial reduction in topside weight, reduced tensioning requirements, elimination of buoyancy elements, surface handling gear and the expensive and cumbersome flex joints both traditionally used with steel drilling risers. Factors inherent in titanium and titanium alloys, such as galvanic interaction, wear and a low threshold for galling had to be addressed and the design solutions incorporated at an early stage.

Costs and the lack of familiarity of offshore engineers and design houses are the main barriers to titanium's wider use on deep water applications. Titanium is not an 'exotic metal, it is relatively inexpensive and widely available and may be readily manipulated, machined and fabricated using conventional practices at little extra cost when compared with high alloy steel. Titanium is as strong as steel, yet 45% lighter. The high strength, low density and corrosion resistance of titanium contribute positively towards cost reduction. All up weight on semi submersible platforms including tension leg platforms is equally critical, the reduction in hang off weight can be matched by 3-5 times weight reduction in the platform structure, flotation and mooring systems. Titanium requires no corrosion allowance so equipment can be designed to satisfy the minimum requirements in mechanical strength and handling. The installation of the Heidrun drilling riser in the first quarter of 1996 and satisfactory performance to date will raise confidence in the offshore industry.

TLP DRILLING RISER DESIGN SYSTEM REQUIREMENTS

The primary function of the drilling riser is to enable the safe transfer of fluids, drilling tools, and completion equipment in conjunction with the rest of the drilling riser system. Internal fluids include sea water, oil and water based muds, sodium chloride brines, hydrocarbons, hydrochloric acid, biocides, methanol/glycol, corrosion and scale inhibitor. Running equipment includes, drill bits, drill pipe, milling tools, fishing and coring tools, under-reamers, packers and safety valves.

The Heidrun drilling riser installation string consists of 24-off standard riser joints, pup joint, tension spool joint, a taper stress joint complete with lower marine riser package, stress joint similar in design to standard riser but with a heavier tapering section at the wellhead/drilling riser flange interface. Spare risers were provided to enable rotation of individual joints after each drilling-retrieval operation, approach being redistribution of applied stresses to each joint.

The design approach for the Conoco/Statoil Heidrun Drilling Riser incorporated a riser system basis of design written by Conoco Norske Inc's Well Systems Engineer's. Each Drilling Riser Standard Joint consisted of compact flanges with an extended weld neck welded to each end of an extruded pipe. The welding of the pipe to the compact flange coupling employed an automated tungsten inert gas (TIG) welding process and Ti-6Al-4V ELI filler wire. A 3mm internal hydrogenated acrylonitrile butapine rubber liner in sheet form, adhesively bonded to the internal surface of the riser provides the required wear resistance. Figure 1 presents the general assembly of the standard riser joint. A boosterline manufactured from ASTM grade 9 titanium is assembled to five off saddle clamps positioned along the outside of the joint and attached to the upper and lower flanges with ancillary titanium fasteners. The saddle clamps also carry the umbilical from the surface to the Lower Marine Riser Package (LMRP) which isolates the riser bore and also controls the function of the wellhead connector.

Figure 1. Assembled Prototype Standard Drilling Riser

The minimum bore of the standard drilling riser is 552.8 mm (22" ins), minimum wall 22.23 mm (0.875" ins). The made-up length measured 14.685 m (48.2 ft). The weight of the joint is 3252 kg (3.2 ton.f) in air and 2421 kg (2.38 ton.f) in water. The water depth on Heidrun is 345 m. Heidrun is the most northerly of TLP's installed to date and has to survive adverse weather conditions. The drilling riser is designed based on loads corresponding to the following design conditions in Table 1.

Table 1. Design Based Load Conditions

Normal Operating	1-Year Storm Field pressure test + 95% non-exceedance environment
Normal Extreme	100-Year Storm Field pressure test + 1 Year Storm
Abnormal	10,000-Year Storm Well Kick + 1 Year Storm

The pressure range is 63 bar (913.5 psi) to 253.5 bar (3675.75 psi) for operating and well kick conditions. The field pressure test is 314.9 bar (4566 psi). The factory acceptance test FAT is 427.6 bar (6200 psi). The bending moment range for the standard joint is 535 to 1515 kN-m. The bending moment for the Taper Stress Joint under normal/extreme conditions is 3655 kN-m. The calculated stresses within the riser pipe are limited to the API RP 2T allowable stress format.

The safety factors on yield strength are as recommended in API 2T for operating, extreme and installation and abnormal design conditions. The design philosophy was developed in accordance with the NPD regulations for pipeline and riser systems. The safety factors for the various design conditions are presented in Table 2.

The joints are designed to 3 times the service life in fatigue. Design Life 15 years and Fatigue Life 45 years. In addition to the design loads the riser design requires that the riser can withstand 250 kJ caused by a dropped object and still have the ability to sustain the applied tension and pressure.

Table 2. Design Safety Factors

Net Section Stress	Design Condition			
	Installation	Normal Operating	Normal Extreme	Abnormal
Yield Strength SF_Y^P	85%	67%	80%	100%
Ultimate Strength SF_U^P	68%	54%	64%	80%

MATERIAL SPECIFICATION

The advantages that titanium and titanium alloys offer over steel for the manufacture of a high pressure drilling riser include low density, high strength, low elastic modulus, corrosion and erosion resistance. All grades of titanium alloy give these advantages ASTM Grade 23 (Extra Low Interstitial Ti-6Al-4V-ELI) in a beta processed or beta annealed condition was selected as offering the optimum combination of strength, fracture toughness and resistance to crack propagation required in the primary components of the riser. Table 3 lists the titanium components and details the alloys used.

Table 3. Titanium alloys used in the Drilling Riser

Riser Component	ASTM Grade/Alloy	Composition	Characteristics
Flange Gasket	Grade 2	Unalloyed Ti	Moderate strength, good ductility and corrosion resistance
Pipes, Flanges, Studs and Nuts	Grade 23 Beta Processed PWHT	Ti-6Al-4V ELI	High Strength & Toughness, (Work Horse alloy)
Attachments & Ancillary Fasteners	Grade 5 Alpha-Beta Processed	Ti-6Al-4V	High Strength availability
Boosterline Pipe	Grade 9	Ti-3.5Al-2.5V	Medium Strength, good manufacturability

The pipe and flange extrusions were the largest ever to be manufactured from titanium within the 'Western Hemisphere'. Table 4 presents the physical and mechanical properties of the titanium alloy and weldments used in the design analysis and manufacture of the primary components.

COMPONENT DESIGN

The Drilling Riser System is presented in Figure 2 and shows the interfaces of the riser joints, titanium primary components, low alloy carbon steel wellhead connector, tensioner spool piece and the lower centraliser bearing.

The design phase was split into titanium and non-titanium components. The initial design and analysis was on the main riser joints, longest lead time items, to secure the production schedule for the pipe and flange forgings. The Lower Marine Riser Package (LMRP) was designed in parallel. The non-titanium parts included the handling equipment, tensioner joint, spider and centraliser bearing. The primary components of the drilling riser comprised the welded joint assemblies consisting of a pipe with upper and lower flanges. A "no loose item" policy was adopted on all parts except for the titanium gasket seal. The mechanical properties validated in the earlier development and prototype programmes were utilised for the pipe and flange design.

Table 4. Mechanical Properties of Primary Component Parts

Minimum Acceptable Properties	Pipe	Flanges	Weld Metal	Studs & Nuts
0.2% Proof Stress Mpa (Ksi)	724 (105)	758 (110)	758 (110)	758 (110)
Ultimate Tensile Mpa Strength (Ksi)	828 (115)	828 (120)	828 (120)	828 (120)
Elongation 5D%	6	5	4	10
Reduction of Area (RA)%	12	10	-	25
Fracture Toughness MPa m (Ksi in)	77 (70)	90 (82)	75 (68.3)	90 (82)

All metal component attachments such as alignment blocks and guide pins of the joints were manufactured in titanium to give maximum benefit of strength to weight ratio for the fully assembled joints, and minimised galvanic interaction. The design took into consideration the limitations of manufacture while meeting the requirements of good engineering practice. It was preferable that intermediate welds on the riser be kept to the absolute minimum - one of pipe with two flanges - giving the minimum of two welds making an overall finished face to face length of 14685 mm. The only exception was the Taper Stress Joint (TSJ) which required a pipe to pipe weld due to the added wall thickness of the pipe. The maximum length that could be extruded on the available presses dictated the length of the flange weld necks.

The advantage of the extended weld neck meant that the weld was at a distance from the flange face such that all secondary bending stresses induced by the make-up assembly and stiffness variation had decayed before exerting any influence on the welded area. A field proven low profile "compact flange" design was fully analysed using finite element analysis, and optimised, to meet the requirements of the operational field specifications and be fit for purpose for use in the TLP drilling riser.

Two-dimensional axisymmetric finite analysis was carried out to investigate the performance of the standard joint compact flange under the static and fatigue load regimes. Analysis was conducted using the general purpose finite element programme ANSYS. The coupling was assessed against the specified design codes in order to verify compliance of the neck profile, studding arrangement, stud preload and mating face and gasket seal profile. In addition, the sealing capacity of the coupling was established for each of the specified load cases. Fatigue assessment of the coupling design showed that all locations receive only minor accumulated damage, and therefore crack initiation is unlikely to occur.

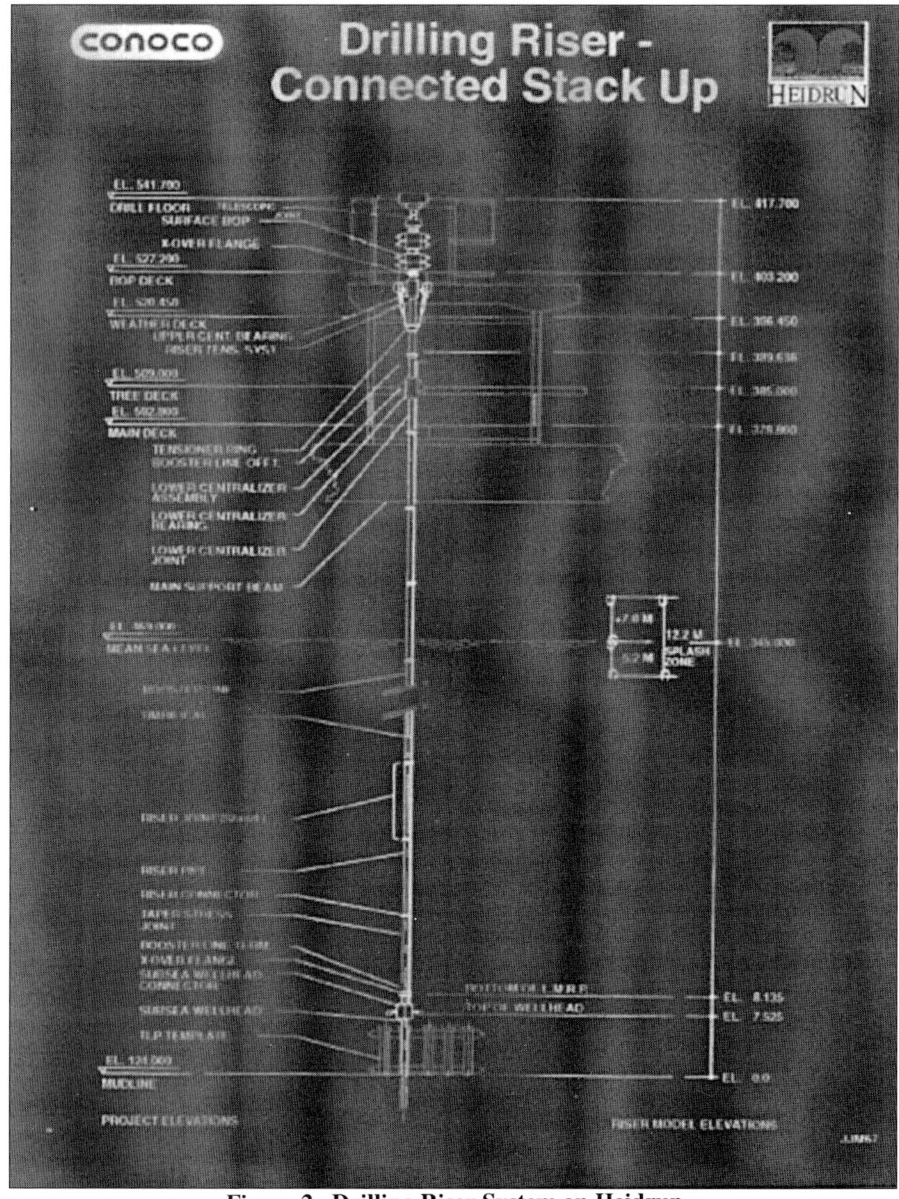

Figure 2. Drilling Riser System on Heidrun

The detailed analysis on the base design of the compact flange highlighted the need to take cognizance of titanium's low modulus of elasticity which is approximately half the value of carbon steel. The modified compact flange design has an inherent high preload and sealing system that is achieved by axial rotation of the two flange bodies during the assembly of the flange. The flange has a conical profile in a convex form accurately machined on both make-up flanges. The small cone angle is reduced due to the convergence of the flange faces caused by the rotation of the bodies on make-up.

The low modulus of elasticity of titanium further effected the design of the flange studs. The studs are tensioned and whilst the extension is applied the nuts are run down the threads and then load released. The greater extension due to the low modulus necessitated that the extended thread pitch be accommodated by the nut thread design, to eliminate jamming. Galling had to be resisted on the nut to stud action and an epoxy polyamide molydisulphide coating was applied to the anodised titanium surface. The coating was applied to all ancillary fasteners.

The booster line pipe, with weld on pin and box stab connectors, was completely manufactured in titanium alloy Grade 9. The boosterline was clamped to the riser joint and supported above the outside diameter of the flange ends. Clamping and support was achieved at the flanges by means of support plates fixed to the back faces of the flanges. U' clamps retained the stab connectors at the flanges, allowing axial freedom of movement between the pin and box mating components, axial movement induced by the flexure of the riser string. Rigid polyurethane moulded clamps and saddles provided intermediate support and clamped the boosterline over the full length of the riser joint. The saddles were secured to the riser pipe by Kevlar tensioning straps.

The tensioner steel–to–titanium flange interfaces required designs to resist galvanic interaction between the face to face contact. Inconel 625 was inlaid in the steel flanges. Neoprene sleeves were used as an isolating barrier between the steel and titanium of the steel centraliser bearing. Where titanium studs were assembled to steel flanges the barrier was achieved by nylon sleeves.

The drilling riser experiences mostly subsea service, but is run and retrieved regularly after drilling and completion of each Well operation. Part of the design strategy was to determine crack growth rate and crack instability under specified long term and extreme loads. The fracture mechanics analysis results were used to determine manufacturing inspection requirements and to establish an in-service inspection strategy for each of the riser components. The minimum inspection interval for drilling risers is 1 year with a design goal of 2-3 years. Visual examination is carried-out after each drilling riser retrieval.

MANUFACTURE

Pipe & Flanges

The pipe and flanges which formed the principal components of each riser joint were produced from triple melted and forged billet stock by hot piercing and forward extrusion. The elongated weld neck configuration required for the flanges was achieved by using an interrupted extrusion stroke in which the 1.5 meter of pipe for the weld neck was extruded with sufficient thickness of material to yield a flange left attached in the extrusion tool.

Figure 3. Pipe and Flange Extrusions

The required equiaxed fully transformed beta structural condition was achieved by extruding pipe from a controlled temperature in the beta field whereas flanges were alpha - beta forged and subsequently beta annealed. Both pipe and flanges received a final anneal at 730°C to assure dimensional and metallurgical stability.

Each pipe and flange component was comprehensively release tested, notably including fracture toughness testing, before going forward for machining. A full volumetric ultrasonic examination was performed on each component at a rough machined stage followed by liquid penetrant examination of all final machined surfaces. Non destructive examination acceptance standards were cross related to limiting defects sizes used in the fracture mechanics analysis referenced later in this paper. A close dimensional control was exercised at all stages of manufacture.

Booster Line Pipe

The small diameter pipe was hot extruded from forged and bored billet stock using extrusion and annealing temperatures in the alpha beta field of the alloy. Finishing of the inner bore was performed by honing and the outer surface by grinding. The relatively small diameter and wall thickness of the pipe limited non destructive testing to visual and liquid penetrant examination.

Ancillary Components

The grade 23 bar stock used for the flange studs and nuts was beta annealed and subsequently annealed at 730°C. Otherwise all ancillary titanium components were machined from standard rolled or forged semi-finished products in an alpha-beta hot worked and annealed condition.

WELDING OF TITANIUM THICK WALLED PIPE

Riser joints were built by welding a flange to each end of a length of pipe using the tungsten inert gas (TIG) welding process. Welding was performed automatically in the 1G position with the stationary head and rotating work piece under rigidly controlled conditions of inert gas shielding, deposition rate and pass sequence, using compatible grade 23 filler wire to AWS ER Ti-5 ELI. The specific parameters used had been previously validated in a fully representative procedure qualification which had been thoroughly examined and tested to the stringent Heidrun welding specification.

A post welding heat treatment at 700°C was applied locally to all completed welds to minimise residual stresses and assure freedom from any detrimental transformation products in the weld bead or HAZ. After being machined and ground back to a smooth inner and outer surface profile with respect to the adjacent pipe and flange, welds were ultrasonically and liquid penetrant examined to the same standards as for the base pipe and flange. A radiographic examination was also performed. Results of non destructive testing were satisfactory and fully justified the care taken in initial procedure qualification and the welding process selected.

The pipe to pipe and pipe to box and pin welds required in the assembly of the booster line were made to a manual TIG procedure with filler wire to Ti-3AI-2.5V (Grade 9). All welds were non destructively examined by radiography and liquid penetrant prior to hydrostatic test of the completed booster line assembly.

S-N FATIGUE & FATIGUE CRACK PROPAGATION

The following is an extract from a paper Titanium Drilling Risers - Applications and Qualification (Salama, et. al., 1999)

S-N FATIGUE

Drilling riser dynamic loads are predominately in the high R-ratio (>0.8) regime. Tensile fatigue test were carried out on base metal and welded joints using dog-bone specimens to validate the proposed design curve for the titanium risers as well as to assess the sensitivity to seawater, with and without cathodic protection. The test specimens were prepared with 3mm radii on the edges to avoid spurious results in testing caused by premature crack initiation. The results of these tests in both air and sea water are presented in figure 4. Overall the S-N data validated the proposed design curve. The results also showed that the effects of seawater with and without CP are negligible. The fatigue test have also demonstrated that the manufacturing defects have a significant influence on crack initiation and therefore on fatigue life. Weld defects result in large scatter of data especially in the lower stress ranges. influence on crack initiation and therefore on fatigue life. Weld defects result in large scatter of data especially in the lower stress ranges.

Nearly all fractures in the fatigue specimens originated from a single defect in each specimen, either intrinsic to the base metal-weld or externally induced (machine marks). The influence of service induced defects on fatigue life may be characterised using the S-N approach in a similar manner as used by the aircraft industry, providing the existence of sufficient data on simulated defects, or can be attempted via crack propagation and fracture mechanics analysis.

The latter approach ignores the crack initiation aspect of the total life. A sample calculation performed on one of the worst S-N data indicated that the crack propagation life is a very small fraction of the total life. Fractography indicated that this defect was the sole origin of the final failure. Assuming the defect to be a crack and applying known crack growth and toughness parameters, the critical crack size was established to be 8.12 mm, which was confirmed by the fractography. However, there was a disparity in the calculated number of cycles for the 0.3 mm "crack" to reach 8.12 mm depth and the actual cycles recorded in the S-N fatigue test.

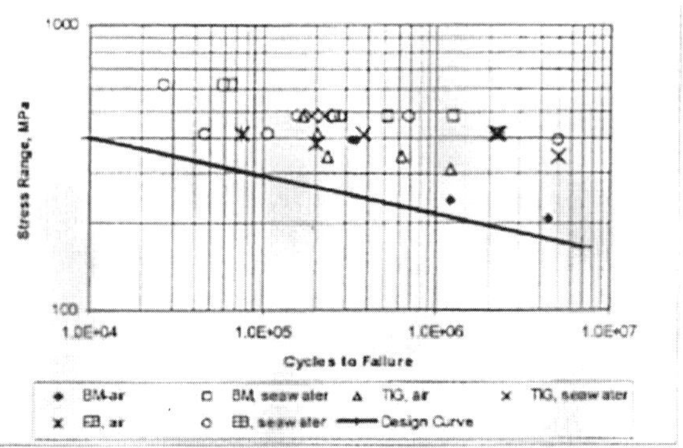

Figure 4. Fatigue S-N Data for T-6Al-4v ELI Alloy in both Air and Seawater

The calculated and actual values were 790 and 45,700 cycles, respectively. This would mean that a mere 1.7% of the life is spent in crack propagation. The same conclusion was reached by analysing other specimens with initial defects. Thus, even for specimens with questionable fatigue life, the disparity is striking, which indicates the extent to which crack initiation mechanisms are operative.

FATIGUE CRACK PROPAGATION

In addition to design life estimates of the drilling riser using S-N fatigue analyses, fatigue crack propagation (FCP) analyses were performed to establish tolerable defect size and inspection frequency. The results of FCP tests on Ti-6AI-4V ELI alloy in air and in seawater are shown in Figure 5. Tests in seawater were conducted at 0.167 Hz.

The results show that stage II crack growth rates in seawater under freely corroding and cathodic protection conditions were slightly higher than in air. The threshold stress intensity values ranged from 1.4 to 3.3 ksi-in$^{1/2}$ (1.5 to 3.6 MPa-m$^{1/2}$). FCP rates were also measured in water based drilling muds. The results indicated that crack growth rates in drilling mud were considerably higher than in air. The FCP data in drilling mud would be primarily applicable to the ID of the riser that is lined, except for areas near the handling groove.

TESTING

Every individual primary component part of the high pressure drilling riser was Factory Approval Tested (FAT) as part of the manufacturing quality plan. All riser joints were pressure tested including the boosterline and LMRP. All lifting and handling equipment was proof load tested. The pressure test on each riser joint was carried out using steel blind flanges for the primary joints and floating pressure end caps for the booster line. Both pressure tests were carried out simultaneously on the fully assembled joints to replicate offshore running conditions.

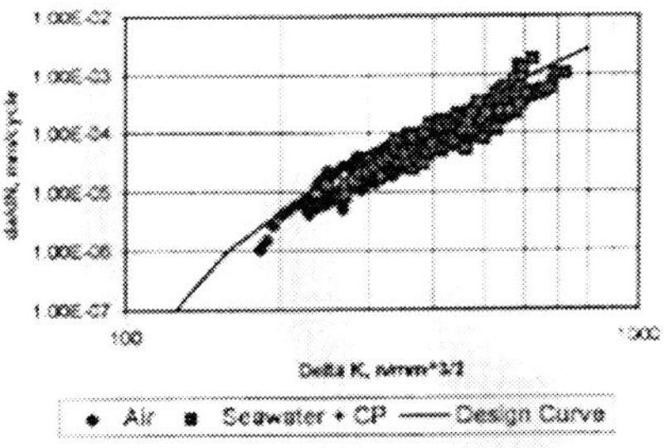

Figure 5. Crack Growth Behaviour of Ti-6Al-4V Alloy & Weldment in Air & Seawaters

Drilling Riser Stack-up Tests were performed to verify the interface and interchangeability of the riser joints. The stack-up trials were conducted at the Downhole Technology Centre, Aberdeen, see Figure 6. The centre provides an independent facility where drilling and down hole equipment operational techniques can be developed, tested and demonstrated without the risk and expense associated with live offshore wells. The facility was chosen to reproduce the stack-up operating conditions as closely as possible to the offshore conditions and provide a safe working environment. The drill floor on the rig was modified, installing a raised platform with opening 49.5" diameter to simulate the rotary table on the Heidrun Platform.

Figure 6. Stack-up Trials at the DTL Land Rig

During the stack-up operation each combination of riser and boosterline was assembled with the flange and boosterline seals fitted. The titanium studs and nuts were tensioned utilising a specially built Bandac Tensioner tool. The tool tensioning 12-off studs during each tension sequence. The assembly and disassembly time and tension loads were recorded for each test.

The specified pre-tension loads and assembled loads were verified by the use of strain gauges attached to the titanium studs. Evaluation of the strain data determined the number of stud tension sequences to be conducted for assembly and disassembly to achieve and relieve the stud pre-load of the assembled flanges. After each flange assembly seal integrity checks were carried out.

The tests also provided performance evaluation of the Drilling Riser Spider and Handling Subs. Stack-up trial assembly was also conducted on the combination of the Taper Stress Joint (TSJ)/Wellhead Connector/LMRP and a Standard Riser Joint. The fully assembled combination was FAT pressure tested via the Wellhead test stump and boosterline by operating the LMRP gate valves, checking the leak integrity of each booster line shut-off valve. The test was undertaken in the horizontal position due to the excessive overall assembled length. TheTSJ/LMRP/Wellhead connector assembly was also run vertically through the 49.5" rotary table, checking for snagging and any potential interference interface problems. Figure 7 shows titanium riser joint on the Heidrun pipe deck.

Boosterline wear tests were performed to demonstrate and verify the capability of the boosterline to withstand the design loads, without any permanent deflection, to observe the wear of the boosterline pipe body, saddle clamps and seal areas of the pin and box stab connectors. A constant bending moment was applied with the pin and box components subjected to axial cyclic fatigue loading. The internal chamber of the boosterline was pressurised (operating pressure) with water based drilling mud. Torsional load tests were further carried out on the saddles to verify the design was fit for purpose.

Stud bolt creep tests were conducted to simulate the flange make-up during offshore running conditions. The stud was pre-tensioned and left in a preloaded condition for fifty days prior to releasing the load. The load test measured the pre-tension load, pre-load and the load transfer loss. The preload was checked at regular intervals throughout the fifty day test. The applied loads were measured by strain gauges and displacement transducers monitored by a data acquisition system. The tests verified that the titanium studs showed no strain variations and that no creep would occur over a 50 day drilling programme under the specified make-up load.

Figure 7. Titanium Drilling Risers on Heidrun Pipe Deck

COMPOSITE DRILLING RISER

Deep water oil and gas exploration and production pose significant challenges that can be addressed by new technologies. Lightweight, high strength composite materials can be viable alternatives to steel or titanium in deep water applications for TLP's, semi submersibles and drill ships. Composite risers weigh less than half the weight of steel and cost significantly less than titanium.

A Joint Industry Project, Compriser, led by Kvaerner Oilfield Products (KOP) and Norske Conoco.AS. (NCAS), supported by Statoil, Shell, Norsk Hydro, Petrobras, Chevron, Saga and EU Thermie has a primary objective by field demonstration to supply a 15 meter long, 558 mm (22 inch), internal diameter, high pressure composite drilling riser joint on the Heidrun Tension Leg Platform (TLP). The composite riser joint is designed to be fully compatible with the titanium drilling riser. Full scale prototypes have been manufactured and will be qualified through rigorous static and dynamic tests. All the design and analysis and test results are verified by DnV. Field trials are scheduled for the first quarter of year 2000.

Oil States Industries UK Ltd, under contract to Kvaerner and Norske Conoco are responsible for the supply of titanium manufactured assemblies to the pre-winding stage. Oil States UKAS-NAMAS accredited test laboratory will conduct mechanical testing of full scale prototypes, assembly and factory acceptance testing of the field joint and the fitment of an internal rubber wear liner. Lincoln Composites, Nabraska, reporting to Kvaerner and Conoco are responsible for the winding of the composite materials.

COMPOSITE RISER DESIGN

The composite drilling riser is designed to be interchangeable with the Standard Titanium Drilling Riser Joints as used currently on the Heidrun Platform. The composite joint has titanium flange configurations similar to the Heidrun joints and is designed to satisfy all dimensional constraints to make the riser suitable for installation and operational use on Heidrun. The advantages of using a composite riser are lightweight, 35-50% of the weight of steel, corrosion resistance, high strength to weight ratio, high fatigue resistance, ability to engineer properties to meet specific design criteria, lower systems costs (weight sensitive structures and lower costs as compared to titanium risers).

The tube body weight of the composite drilling riser is 67% of an equivalent titanium joint. The fatigue life of the composite riser joint exceeds 150 years, 10 times the service life. The internal pressure rating exceeds 12,600 psi (prototype test sample to failure test burst at 15,850 psi), which is 3.5 times the maximum operating pressure 4500 psi of the titanium riser. The composite drilling riser without protection maintains its pressure and structural integrity after being subjected to a dropped 13.3/8" drill pipe, impact of 50 kJ.

The structural capability of the composite drilling riser tube body is provided by a carbon fibre-epoxy composite overwrap, with load transfer between the composite overwrap and titanium flange extensions accomplished through designed traplock metal-to-composite interface (MCI), see figure 8. Detailed FEA analysis of the composite riser confirmed that the riser is capable of safely supporting all the design loadings, including internal and external pressures, axial tension, bending and impact loads. The bore of the composite drilling riser joint is provided with a damage resistance and leak resistance barrier, titanium liner, material Grade 9, wall thickness 0.125" (3.2mm), plasma welded to the Grade 9 transition ring of the titanium MCI/Flange. An elastomeric rubber liner is bonded to the internal bore of the drilling riser to protect the titanium from wear. The rubber liner is identical to the current liners used on the Heidrun titanium drilling risers.

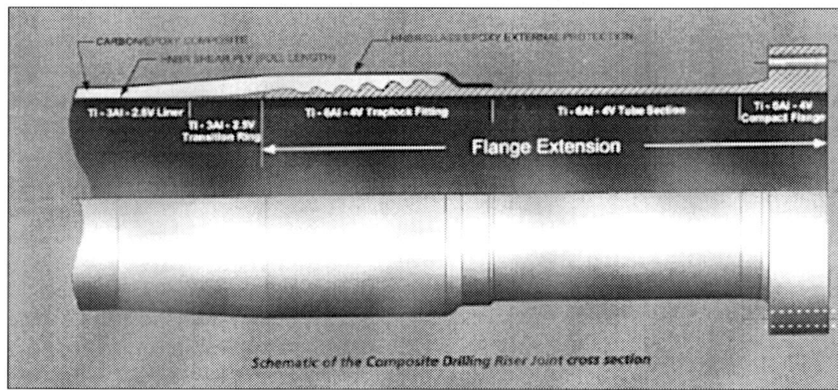

Figure 8. Metal to Composite Interface (MCI)

The Compriser project is being conducted in three phases. Phase 1 design, engineering and qualification programme which includes static and dynamic testing. Testing will include burst, combined pressure and bending, fatigue loading as well as impact resistance evaluation of protection systems. Phase 2 is to fabricate a 15 metre full length riser joint complete with ancillary attachment saddle supports and booster line. Phase 3 will involve field testing of the composite drilling riser. The composite drilling riser will be installed in three separate drilling operations. The riser positioned in various positions 50 metres below the splash zone.

Qualification testing Phase 1 included impact trials on bare laminate pipe to impact range 25-50 kJ. Impact test of laminate pipe with protective jacket, impact 250 kJ. Test Sample 1 was successfully burst to failure, see Figure 9. Test Sample 2 will be submitted to 4 point bend cyclic fatigue with internal pressure and service temperature, then finally subjected to maximum bend to achieve design safety factor. Test Sample 3 will be subjected to 4 point bend cyclic fatigue similar to Sample 2 but 10 times service life. Figures 10 and 11 show test samples 2 and 3 prior to composite overwrap winding. Figure 12 presents the field test sample prior to winding. Trials are ongoing and are due for completion December 1999. Field joint trials are scheduled for first quarter year 2000.

SUMMARY

Advanced composite materials offer unique properties such as light weight, high strength, good thermal insulation and excellent fatigue and corrosion resistance, which make composites attractive to deep water applications. The successful application of composite risers require key technical issues to be addressed which are unique to the composite components, design, manufacture and performance.

Figure 9. Burst Test Sample 1

Figure 10. Test Sample 2 with Composite Winding

Figure 11. Test Sample 3 Prior to Composite Winding

Figure 12. Field Joint Prior to Composite Winding

Projects currently in progress to qualify composite production risers and drilling risers are the aforementioned Heidrun composite drilling riser, the NIST-ATP (US Department of Commerce National Institute of Standards & Technology - Advanced Technology Program) project on composite production risers and the Norske Conoco A/S and Oil States Composite Steel-Titanium Interface connector project.

The Heidrun titanium drilling riser has been in operational use for 3 years. Drilling operations have seen the successful completion of 15 new wells. The risers are inspected after each retrieval, random checks are conducted on the titanium riser welds, pipe material and titanium studs by industry approved NDE methods. The titanium materials have shown no detrimental defects. Periodic repairs have been carried out on the internal rubber wear liner. Damage mainly due to initial operational start-up procedures, revised drilling procedures have seen limited damage to the rubber liner. Booster line pins have seen wear caused by abrasion, mud being trapped between the pin and box elastomeric seals. This operational damage has been rectified by coating the titanium pin surface with Tungsten frame spray technique.

Although titanium Ti-6Al-4V ELI alloy offers the optimum combination of elastic modulus, strength and weight for drilling riser applications, titanium is quite expensive. Therefore efforts are underway to evaluate carbon fibre composites as an alternative solution. The proposed design of composite joints incorporating titanium liner and connectors suggest that composite drilling risers will cost approximately 50% than that of titanium risers.

ACKNOWLEDGEMENTS

The author gratefully acknowledges input by Norman Lumsden of Kongsberg, Dr Mike Smith of Oil States, Mamdouh Salama and Jay Murali of Conoco, and Lincoln Composites, Nebraska.

REFERENCES

American Petroleum Institute (1997), "Recommended Practice of Planning, Designing and Constructing Tension Leg Platforms", API Recommended Practice 2T Second Edition.

Salama, M.M., Echtermeyer, A., Linderfjeld, O., "Composite Risers for Deepwater Applications".

Thornton, J.M., "The Heidrun TLP Titanium Drilling Riser".

Salama, M.M., Murali, J., Joosten, M.W., "Titanium Drilling Risers – Application & Qualification".

Lumsden, N., Panayotti, A., "Titanium Drilling Risers – Design, Manufacture & Test".

S696/006/99

Titanium in architecture – a consulting engineer's view

S P CARDWELL
Ove Arup & Partners, London, UK

1. INTRODUCTION

The October 1997 opening of the Guggenheim III Museum, Bilbao, Spain, arguably inspired the current growth in the wider interest in the use of titanium sheeting for roofing and cladding applications in Northern American and European architectural circles. Frank O'Gehry, the Californian based architect, designed the building which was described by Building Magazine as *'Frank Gehry's latest abstract sculpture'* [Ref 1] opened to wide critical acclaim (Figure 1). In addition to the complex and amazing three-dimensional structural form, which was designed and modelled using software more conventionally used by designers of military aircraft, the other significant feature of the building was the metallic cladding.

Originally O'Gehry had intended to use 'lead-copper' sheeting. However, this was outlawed by local authorities as a toxic material. Stainless steel was evaluated as an alternative, but this was thought to be too 'cold and industrial'. During this period of the design process they found some samples of titanium which lead them to 'realise there was some potential for a metal that had warmth and character' [Ref 2]. This realisation ultimately resulted in sheet titanium being selected and specified for the cladding of the museum. The museum is actually clad in 33,000 separate, but interlocking panels of 0.4mm thick commercially pure sheet titanium, covering a total surface area of 35,000m^2.

The high profile of the Guggenheim Museum has certainly raised the interest in titanium beyond all recognition. However, contrary to popular opinion, the museum is not the first example of the use of titanium in architecture; far from it. The material has been used for roofing and cladding applications by the Japanese since the 1970's where there are now in excess of 300 buildings of varying size and function.

In this context, this paper will provide a critical review of the use of titanium in architecture. It will question whether the current use of the material is representative of true architectural and engineering innovation or whether it is a fashionable trend.

2. THE JAPANESE EXPERIENCE

Whilst the Guggenheim may well have raised the profile of titanium on a world scale, the credit for pioneering the use of titanium in this application must rightly be given to the Japanese. They recognised the potential to utilise titanium in order to address atmospheric corrosion problems and the associated reduction in durability of other metallic construction materials.

The very first example dates to 1973 where a modest 150m^2 of 0.4mm sheet was used to roof the Hayasuihime Shrine, Saganoseki City-Oita. The shrine is located in an aggressive marine environment for which the benefits of improved corrosion resistance are self-evident. The project also represented another significant first. The titanium sheeting was anodised gold. This is only one of a multitude of different coloured finishes that are possible.

The use of titanium in Japanese architecture has grown from this modest beginning to the point where there are now in excess of 300 existing examples. Titanium has traditionally been regarded by the construction industry as being exotic and, more importantly, expensive. As such, it might be expected that the material would only be used for high profile, prestigious projects with budgets to match. However, data published by the Japan Titanium Society indicates this is not the case [Ref 3]. The diversity of projects for which the metal has been used is somewhat surprising. Examples include private housing, schools, public libraries, as well as the more obvious office developments, museums and sport facilities.

There are too many Japanese examples where titanium has been used in architecture to review individually. However, one of the most striking and spectacular buildings that does merit individual mention is the Fukuoka Dome, Fukuoka. The Dome, a sports facility has an 'iris' type retractable roof (Figure 2). This project utilises 42,000m^2 of 0.3mm sheet dull finished titanium.

In addition to the building applications described above, titanium has also been used in Japan for Civil Engineering projects. The best example of which is the Trans-Tokyo Bay Highway Bridge (Figure 3). The bridge piers are fabricated from normal structural steel. The splash and tidal zone of such structures are conventionally protected using a combination of a cathodic protection system in conjunction with a high build paint system. However, this approach to corrosion protection offers a limited durability and requires regular maintenance throughout the life of the structure, which for a bridge can be assumed to be in excess of 100 years. In order to provide a 100 year maintenance free life, titanium-clad steel plate was developed for the application and incorporated into the structure.

Titanium–clad steel is an excellent example of an appropriate application of a new material; the composite plate combines the strength of the base steel plate with the superior corrosion resistance of the titanium cladding. The titanium cladding represents between 10 and 20% of the overall thickness of the plate. This represents a marked cost saving compared with the use of a single thicker section of titanium. It should also be noted that dissimilar metal welds between steel and titanium are not possible, as the titanium becomes embrittled. As such, only steel-steel connections are possible.

3. WHY TITANIUM?

As noted above, titanium has been and still is regarded by the construction industry as being both exotic and expensive. There are many metals that have and will continue to be used successfully for cladding and roofing applications, for example stainless steel, aluminium or copper alloys. As such, the inevitable question that must be asked is why titanium?

A number of the fundamental reasons for choosing titanium have been mentioned above through consideration of the Japanese experience. The following section will expand upon these points and discusses whether the perceived advantage is genuine.

3.1 Corrosion Resistance

Titanium is a very reactive metal. So much so that when the metal surface is exposed to air, or any other oxygen containing environments, the metal reacts with the available oxygen to form a tenacious oxide film. The presence of the oxide film on the surface protects the metal from further reaction and degradation. It is the stability of this oxide layer in a variety of environments that confers the excellent general corrosion resistance.

Fundamentally, corrosion resistance is without question one of the key attributes that resulted in the material being adopted by the Japanese construction industry. Titanium is essentially immune to corrosion in industrial and marine environments. This applies even in climates with elevated levels of relative humidity; a condition which leaves most other metallic construction materials wanting. Additionally, the Japanese have some very specific and unusually aggressive environmental conditions (volcanic fallout and hot-springs) which necessitates very careful consideration with respect to material selection and durability.

The need to specify an expensive, corrosion resistant material such as titanium must reflect the environmental service conditions. In Europe, for example, the environmental conditions are in generally less severe than might be encountered in Japan. As such, the need to specify titanium purely on corrosion resistance is considered not proven. There are many examples where other metallic materials have been satisfactorily used for similar applications.

Cladding systems often use a combination of different metals in direct contact. This can give rise to the potential for additional bimetallic corrosion. Bimetallic corrosion occurs when two different metals are in electrical contact and are also bridged by a common electrolyte, usually water. Current flows from the anodic or baser metal to the cathodic or nobler metal. As a result, the cathodic metal tends to be protected and the anodic metal may suffer additional corrosion.

In the case of the architectural use of titanium, the metal couple that is most likely to be encountered in practice is that of titanium and austenitic stainless steel, where the stainless steel is likely to be used in the manufacture of fixings and fasteners. In most environments, and certainly those likely to be encountered in cladding applications, the electrochemical potentials of stainless steel and titanium are similar. As such, bimetallic corrosion is unlikely.

This is supported in practice where the couple between titanium and austenitic stainless steel has not resulted in any additional corrosion.

The absolute confidence in the corrosion resistance of titanium has lead to at least one of the leading material suppliers to provide a 100 year warranty against corrosion. This is a very laudable gesture. However, it should be remembered that the titanium is but one element of a cladding system. It is the performance of the system as whole that is of paramount importance. This raises this issue of appropriate design which will be discussed later in this paper.

3.2 Physical properties

Table 1 details the physical properties of titanium and other common metals used in roofing and cladding applications.

Table 1. Comparison of physical properties

Physical Property	Metal			
	Titanium[1]	Stainless steel (grade 316L)	Aluminium	Copper
Density (kg/m^3)	4,500	8,000	2,710	8,900
Young's Modulus (kg/mm^2)	10,850	20,400	7,000	11,000
Coefficient of linear expansion (cm/cm/°C)	8.9×10^{-6}	16.5×10^{-6}	23×10^{-6}	17×10^{-6}

It has been argued that the reduced density of titanium can be utilised to reduce the size of the supporting structural frame. However, there is little evidence to suggest this potential benefit has been used in practice.

The stiffness of titanium is approximately half that of stainless steel. Thus, titanium is readily formed using conventional fabrication techniques. However, the spring-back of titanium is greater than that of stainless steel and should be considered during fabrication.

The low thermal coefficient of expansion is considered to be a significant advantage. It reduces the need for numerous expansion joints so that long span roofs are more easily achieved. An example of which is the Kawasaki Citizens Museum, Kawaski-Kanagawa, Japan, which has spans up to 47m in a seam welded construction without the use of expansion joints [Ref 4].

3.3 Fire resistance

Like most other metals, titanium is non-combustible and satisfies standard fire resistance requirements, such as spread of flame for example, and requires no special precautions.

---

[1] Reference to titanium in an architectural context for the present relates to commercial purity material, i.e. 99% purity, and not an alloy of titanium.

3.4 Environmental impact

Environmental impact, or more specifically sustainability, is increasingly being brought to the fore in architecture. Although the term remains ill defined, three distinct, but inter-related issues seem to be coming to the fore: life cycle cost, embodied energy, and appropriate design.

Assuming there is a clear technical need for the use of titanium, i.e. the environment is sufficiently aggressive such that other materials may suffer premature degradation, as has been demonstrated in Japan, life cycle costing can be used to demonstrate the benefit of specifying titanium.

The immediate competitor of sheet titanium in Japan is PVF_2 coated stainless steel. Life cycle costs data published by the Japan Titanium Society indicates that the initial cost of titanium is approximately twice that of the coated stainless steel [Ref 3]. However, when consideration is given to the maintenance costs associated with the additional corrosion protection used in conjunction with the stainless steel, the costs of the two materials are comparable after 20 years. Thus, titanium offers a potential cost saving for buildings with a projected design life in excess of 20 years. Additionally, titanium is totally recyclable should the need arise.

Despite the fact that titanium is one of the most abundant elements on the planet, the high initial cost of titanium is a reflection of the high energy demands for extracting the metal from its ore and the subsequent processing. Thus, embodied energy associated with sheet titanium must be regarded as being high. Although, this appears to be a significant disadvantage, consideration must be given to the life cycle of the building.

Appropriate design is really a process of finding a design solution which recognises the balance between design needs and desires, resulting in appropriate material selection and design details to achieve them. For example, whilst titanium has an undeniable excellent durability, the benefit of this could be reduced by the inappropriate choice of materials within the cladding system or the method of fixing, both of which could affect the long term performance of the cladding system as a whole. As such, detailing is of prime importance. Furthermore, such problems may not be immediately apparent.

Environmental impact must not be confused with simply being environmentally safe. Given the inherent corrosion resistance of titanium, no metal ions are dissolved by rainwater and, therefore, there is no associated risk of environmental contamination. Furthermore, the use of titanium in medical science, for prosthesis for example, clearly demonstrates the non-toxicity of the metal to humans.

3.5 Fabrication Characteristics

It is beyond the scope of this paper to discuss in detail the various fabrication techniques that are utilised in the manufacture of titanium panels for architectural applications. However, it should be noted that titanium is readily fabricated using all of the common techniques that are utilised for stainless steel. As such, although titanium is a new material to many of the cladding manufacturers, its use should not create any fabrication difficulties.

3.6 Aesthetics

All of the reasons discussed for using titanium thus far have been of a technical nature. However, one must not forget artistic requirements. In certain circumstance these can be as, if not more, important that purely technical arguments. Simply the desire to do something different is a justified reason to use titanium in its own right. As noted it was the aesthetic of a small sample which motivated Gehry to adopt the metal for the Guggenheim Museum.

There are a number of different commercially available finishes:

- standard mill finishes (ranging from dull through to bright finish)
- polished and brushed finishes (ranging from hairline to mirror finish)
- shot blast finishes
- embossed finishes
- colour anodized finishes

The colour finishes are achieved using an anodizing technique that alters the thickness of the natural oxide layer. As the thickness of the oxide layer increases light interference effects, illustrated in Figure 4, result in an apparent colour change. The process itself is relatively simple. The metal is dipped into an electrolyte and a direct current is passed through the titanium and electrolyte. As the voltage is increased the oxide layer increases thus altering the colour. Most colours are achievable with the exception of white, black and red. Additionally, multicoloured effects are equally achievable through the use of masking techniques and by applying colours associated with the higher voltages first. This is because that once formed, the lower voltage will not affect the colour produced by the higher voltage.

The author is aware of some light stability problems that have been encountered with mill finished titanium in Japan, although details are not readily available. It is understood that in certain circumstance the mill finish has taken either a light blue of light gold hue. This implies a slight change in the thickness of the oxide layer. It is understood that research is being conducted to establish the cause of this effect, but as yet there are no conclusive findings.

4. FASHIONABLE INNOVATION?

This paper has reviewed the use of titanium in architectural applications. It has provided a brief historical review and discussed the reasoning behind the adoption of titanium in architecture. By way of conclusion, the following section will discuss whether the current application of titanium in architecture is representative of true innovation or whether it is a fashionable trend.

In the author's opinion, true innovation with respect to the use of a new material is realised when the application of the material is fundamentally unique, i.e. the design solution is dependant upon a unique attribute of the material and, as such, can not be realised by another similar material. For example, shape memory alloys that exhibit a unique phase transformation, which is triggered by a change in temperature, have been used in the manufacture of actuators for fire dampers. This design solution could not have been achieved using any other metal.

In this context, the use of titanium in architecture is considered to be paradoxical. Titanium has thus far only been used as a replacement for other roofing and cladding materials. Therefore, one might conclude that this application is not representative of true innovation. However, where the environment is sufficiently aggressive, corrosion free performance may only be achieved through the adoption of titanium. Thus, in this instance one might conclude the opposite. This paradox conveniently illustrates the concept of appropriate design and/or innovation and clearly demonstrates its context sensitivity.

In conclusion, there is no right or wrong answer to the question is the current use of titanium in architect representative of innovation or fashion. In many respects both are equally valid reasons for the its adoption, provided the design solution is appropriate.

REFERENCES

1. Wow in Bilbao. Building, 27 June 1997.
2. Frank O'Gehry, Guggenheim Museum Bilbao. Coosje Van Bruggen. Guggenheim Museum Publications, 1997.
3. Reference Guide to Building Material of Titanium – Roofing. The Japan Titanium Society.
4. Titanium in Architectural Applications. The Japan Titanium Society, 5 October 1994.

Figure 1. Guggenheim Museum, Bilbao, Spain (reproduced from Timet trade literature).

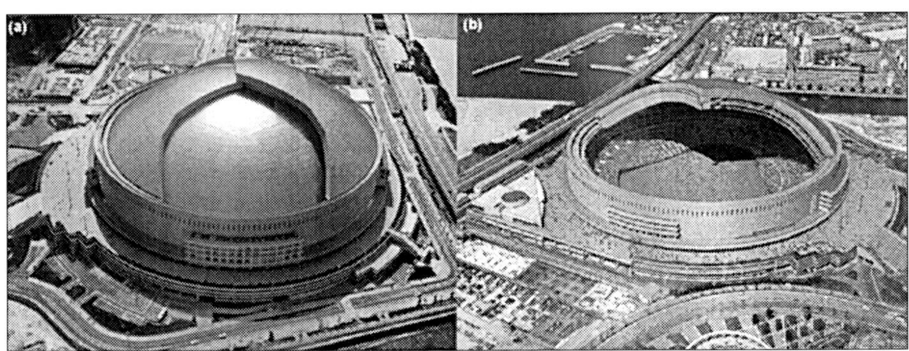

Figure 2. Fukuoka Dome, Fukuoka, Japan – Retractable Roof (reproduced from Nippon Steel trade literature).

Figure 3. Trans-Tokyo Bay Highway, Japan – Titanium-clad Steel Plate Bridge Piers (reproduced from Nippon Steel Corporation trade literature).

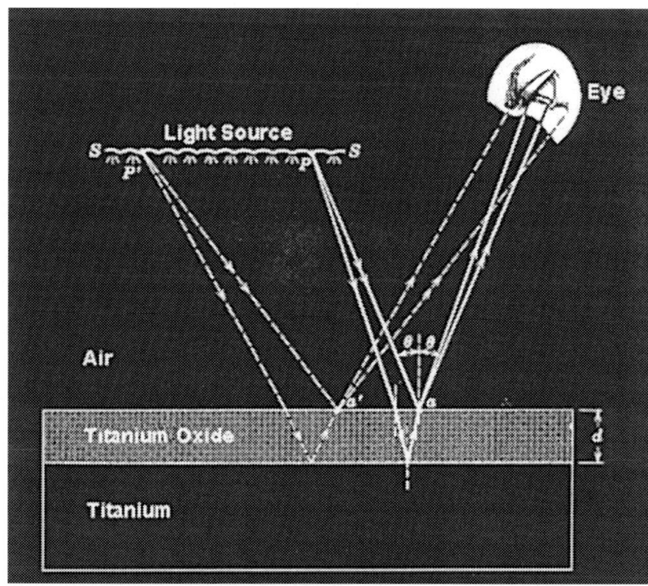

Figure 4. Schematic of the light interference mechanism of colour anodized titanium (reproduced and modified from an original shown in Physics, Part I and II combined, 3rd Edition. Halliday.D and Reswick.R. Published 1978).

S696/007/99

Fibre-reinforced titanium – technology and applications

J G ROBERTSON
Formerly DERA, Pyestock, UK

ABSTRACT

Fibre reinforced titanium offers dramatically improved specific stiffness and strength over conventional monolithic titanium alloys. These properties can be exploited using innovative design solutions for leading edge applications. This paper describes the manufacture and design of affordable selectively reinforced titanium composite components.

1. INTRODUCTION

The advantages of using high strength, high stiffness fibres in polymer composite materials has been demonstrated in many applications. The idea of reinforcing metals with fibres is not new, however, it is only in recent years that suitable fibres have become available. The techniques for incorporating such fibres in real structural parts have been the subject of intensive study in the US, Europe and Japan, particularly for aeroengine applications [1]. Unlike monolithic metallic materials, fibre reinforced metals cannot be forged or extruded since fibre damage is likely to occur. Components must therefore be fabricated to shape. When titanium is used as a matrix there is the additional problem that when molten it is highly reactive and will destroy all known fibres. Fortunately, solid state techniques can be used which allow complex structures to be fabricated utilising the diffusion bonding capability of titanium.

2. MECHANICAL PERFORMANCE

Figure 1 shows how the strength and stiffness of 35% fibre reinforced titanium composite compares with more conventional monolithic engineering materials. They owe these outstanding properties to the silicon carbide reinforcing fibres or monofilaments that are very stiff (400 GPa) and strong (3500 MPa).

Table 1. Properties of SiC monofilament, Ti-6Al-4V and Composite.

	UTS (MPa)	Modulus (GPa)
SiC Monofilament	3500	400
Ti 6-4	1050	100 – 115
Ti-6-4 + 35% SiC	1750	200

Unlike polymer composites, titanium metal matrix composites (TiMMCs) maintain their good mechanical properties in elevated temperature applications. Depending on the matrix alloy, they can be used at temperatures up to 600°C where they have good low cycle fatigue and creep performance. The fibre matrix interface is engineered to prevent degradation of the fibre during processing and to give optimum fatigue crack resistance. At certain stress levels, intact fibres can bridge growing fatigue cracks reducing the stress intensity at the crack tip and reducing the crack growth rate. The transverse strength of the composite (typically 500MPa) is lower than the longitudinal strength but is much higher than the transverse strength of polymer composites. The monofilaments are much larger in diameter than the fibres used in polymer composites. At 100-140um, they are less prone to micro-buckling and therefore TiMMC is particularly good under compressive loads with a strength of approximately 3000 MPa. With care, TiMMC structures can be designed using selective uniaxial reinforcement, positioning the fibres only in regions where component stresses are high.

The main draw back of the material is the cost of designing and fabricating components. Monofilament is currently expensive, composite fabrication techniques are labour intensive and component design is complex. Cost reduction is therefore a key development target. However, in some applications, it has already been demonstrated that the high cost of TiMMC can be justified. Atlantic Research Corporation (ARC) recently published a set of price goals for various aerospace applications [2]. ARC claim that for some applications, like military aeroengine links for example, each pound of TiMMC used will save nearly 3 lbs of component weight. Thus even at $2350 /lb, the use of TiMMC can be justified in this application.

Table 2. ARC established target prices for TiMMC applications [2].

Customer type	Application	Weight savings premium	TiMMC price to achieve weight premium
Military	Pistons	$520 /lb	$1070 /lb
Military	Links	$825 /lb	$2350 /lb
Military	Rotors	$2000 /lb	$1500 /lb
Military	Shafts	$1500 /lb	$1000 /lb
Military	Airframe	$500 /lb	$1000 /lb
Commercial	Airframe	$300 /lb	$800 /lb

3. MONOFILAMENT PRODUCTION

Silicon carbide monofilament is made by chemical vapour deposition (CVD), in a process is shown schematically in Figure 2. An electrically heated tungsten or carbon fibre substrate is passed through a glass reactor containing process gas. Silicon and carbon containing chemicals (organochlorosilanes) react on the hot substrate to form silicon carbide. The fibre is spooled at constant speed from the bottom of the reactor. Un-reacted process chemicals are recycled, reducing feed stock and disposal costs. The length of fibre produced is normally only limited by the length of the substrate so it is most cost effective to operate reactors 24 hours a day, until the end of the substrate. However, since the demand for fibre is limited at the moment, 24-hour operation can only be justified in short bursts. A large proportion of the cost is labour associated with setting up and supervising reactors, unfortunately, at the current low production volume it is not yet cost effective to further mechanise the process. Ultimately, unmanned operation is planned. Figure 3 shows a cost/volume curve for intermediate production quantities. At larger production volumes the cost is forecast to come down significantly, however, it is not envisaged that the fibre will ever become a cheap commodity.

4. COMPOSITE FABRICATION

A number of different composite fabrication routes have been developed for TiMMC components. They involve the use of titanium as either foil, wire, powder or as a coating deposited on to the outside of the monofilament [3]. Figure 4 schematically shows the structure of each process as the component is laid up. If implemented correctly all of these techniques are capable of producing high quality composites with regular fibre spacing. In practice, the fabrication route of choice depends on the shape of the intended component. An important factor that needs to be considered when determining a manufacturing route for a component is the amount of void removal or 'de-bulking' that occurs during processing (Figure 5).

In certain component shapes excessive void removal can place strain on the fibres during consolidation leading ultimately to fibre breakage. For flat panel or aerofoil type structures virtually all the void removal occurs perpendicularly to the plane of the fibres, hence manufacturing routes with high de-bulking such as foil-fibre can be used without damaging the fibres. For radially consolidated rings or shafts void removal becomes more difficult and manufacturing routes with minimal de-bulking such as matrix coated fibre are preferred. Matrix coated fibre has the additional advantage of guaranteeing no touching fibre in the final composite thus producing the highest quality composite. The wire route is also suited to ring shapes or long thin uniaxial structures however there is no guarantee that a good fibre distribution can be maintained. There is a balance between process cost and component performance requirements in determining which route is chosen. Matrix coated fibre probably offers the highest quality product but powder will probably provide the lowest cost solution for many shapes. The cost of pursuing all of these routes means that manufacturers have tended to specialise on their favourite technique so opinions vary as to which will become the dominate fabrication method.

DERA is working mainly on the matrix coated fibre route to high quality components [4]. Figure 6 shows diagram of the process. Titanium alloy rods are melted and evaporated from a water-cooled copper crucible using a powerful electron beam. An array of moving monofilaments is uniformly coated as it passes through the vapour plume. The fibre then leaves the process chamber though an air to vacuum seal. A laser measuring system is used to monitor and control the fibre diameter ensuring that a uniform, smooth coating approximately 50μm thick is obtained. Figure 7 shows micrographs of the as-produced matrix coated fibre and of the microstructure of a composite consolidated from MCF.

A big advantage when using matrix-coated fibre is that it is amenable to filament winding. Modified polymer composite technology can be used to accurately produce intermediates and component structures. The most cost-effective methods involve filament winding MCF directly into titanium tooling which then forms part of the component. Alternatively, flat sheets or aligned fibre arrays can be used as intermediates in the production of shapes not suitable for direct filament winding. The next stage in the process is to degas the assembly and seal it ready for hot Isostatic Pressing (HIPping). High gas pressure is applied to the outside of the component at the diffusion bonding temperature. Under these conditions, not only is the composite fully consolidated but high quality parent metal strength joints can be made between parts of the tooling thus allowing innovative cost saving designs to be employed. Superplastic forming and diffusion bonding (SPF/DB) have been combined for use in a number of selectively reinforced parts to produce near net shape structures in a single step process. Once consolidated, the monolithic titanium can be finish machined to allow component loads to be transferred into the reinforced parts of the structure. The process will now be described in more detail with reference to three example applications.

5. APPLICATIONS

The following examples briefly describe how the unique properties of TiMMC's, coupled with innovative design methodologies can be used to produce components with significant performance advantages.

5.1 Bladed Ring

Figure 8 shows how conventional aero gas turbine engines use high strength discs to support rotating compressor blades. In conventional discs, the material associated with the blade fixing points adds mass to the disc perimeter which turn requires support from a substantial hub and relatively thick disc. In more advanced designs, the blades are directly attached to the disc to form a bladed disc or "blisc", reducing the mass needed for the fixing and allowing weight to be removed from the hub and disc. However, TiMMC is strong and stiff enough not to need a disc to support the high centrifugal loads, even at the operating temperatures within the compressor. Weight savings of up to 70% have been forecast using this type of bladed ring or "bling" design. Additional knock-on weight and cost savings can be made in bearings, shafts and casings.

Conventional titanium alloy discs must be meticulously forged and heat-treated to maximise component low cycle fatigue performance. These expensive forgings require substantial machining, generating large quantities of scrap titanium. Near net shape ring rolled preforms can be used to make TiMMC blings since the performance of the structure is determined more by the performance of the composite than the microstructure of the unreinforced titanium regions. The fibre can be wound directly into a titanium preform and then hot isostatic pressing is used to consolidate the structure. Compressor blades can then be friction-welded directly onto the composite ring. This method significantly reduces the machining time, reduces the quantity of scrap titanium, and thereby reduces the cost of the final component.

5.2 Composite shafts

Polymer composite drive shafts can provide a lightweight alternative to conventional metallic components. Properties such as dynamic instability or 'whirl' in a shaft are dependant upon the shaft dimensions and the specific stiffness (E/ρ) of the material used to make the shaft Most 'engineering materials' i.e. steels, aluminium alloys and titanium alloys have specific stiffness values of around 25, therefore when designing a shaft for increased whirl tolerance changing the material that the shaft is made from has little effect [5]. Composite materials in general have higher specific stiffness values and therefore offer performance improvements. Polymer composite shafts can be used in low temperature applications, for high temperature applications MMC's may be used.

In a gas turbine, using TiMMC means that for a given shaft design the shaft can run faster before dynamic instability occurs. Alternatively, for a given whirl performance the diameter of the shaft can be reduced allowing reduced compressor disc bore stresses and the use of smaller bearing packs. Other issues to be considered include designing compact end fittings to transfer external load to the efficient composite torque carrying structure. In other applications, high temperatures combined with the use of aggressive lubricants limit the use of polymer composite shafts. Fibre reinforced titanium shafts overcome these issues since they are resistant to most lubricants, they are capable of operation at high temperature and can be designed to have much stronger, more compact end fittings. For example, by using an interlayer to maximise joint performance, dissimilar metal spline attachments can be employed. The basic composite torque transmitting structure can also be optimised by varying the fibre architecture to meet specific torque, stiffness or whirl requirements.

5.3 High strength titanium bolts

Selectively reinforced bolts and push rods offer a solution in niche engineering applications where high tensile and compressive strength is needed at reduced weight. Uniaxial fibre reinforcement runs along the length of the component, which is loaded via the unreinforced cladding that surrounds the composite.

Figure 9 shows examples of a valve stem and an engine bolt. Here, the composite has been made by pultruding aligned matrix coated fibres into a hollow titanium rod before HIPping to consolidate. In such applications the accurate control of the coating thickness and chemistry is not as critical as in the compressor bling described earlier. Lower cost techniques for coating the fibre with molten or powdered titanium are being developed. If successful these will open the way for more widespread use of TiMMCs in engineering applications where the current high cost cannot be justified.

6. SUMMARY

It is unlikely that TiMMC will ever be considered as low cost materials due to the complexity of silicon carbide monofilament production and the difficulty in fabricating components. However when production volumes increase a significant cost reduction will be realised.

Three examples have been described which utilise:

- the material's high temperature low cycle fatigue performance,
- its outstanding specific stiffness,
- and its unique uniaxial properties.

The choice of the most cost effective fabrication route will be dependent on the component and application. Lower cost methods for incorporating the titanium matrix are available but most applications are likely to have very demanding performance requirements that can only be met by high quality component manufacturing techniques such as the matrix coated fibre method.

TiMMC components must be manufactured to shape. The use of near net shape fabrication methods such as SPF/DB and ring rolling should help to minimise unnecessary machining and reduce scrap generation.

Selective reinforcement, using fibres only where they are most needed, allows fibre costs to be kept to a minimum and allows any machining operations to be kept within the unreinforced regions. The use of innovative design and manufacturing techniques should therefore make TiMMC cost effective in leading edge engineering applications.

7. ACKNOWLEDGEMENT

British Crown Copyright 2000/DERA. Published with the permission of the controller of Her Britannic Majesty's Stationary Office.

8. REFERENCES

1. Kandebo S.W., Aviation Week & Space Technology, pp156–160, June 14 1999.

2. Hanusiak, W.M., "Titanium Matrix Composites Affordable Manufacturing". Presented at Aerospace Materials 1999 Toulouse, September 1999.

3. Ward-Close, C.M. and Robertson J.G., "Advances in the Fabrication of Titanium Based Composites". Advanced Performance Materials 3, 251–262 (1996).

4. Ward-Close, C.M. and Partridge, P.G., *J. of Mat. Sci.*, 1990, **25**, pp. 4315–4323.

5. Clyne, T.W. and Withers P.J "An Introduction to Metal Matrix Composites" Cambridge University Press 1993 ISBN 0 521 41808 9.

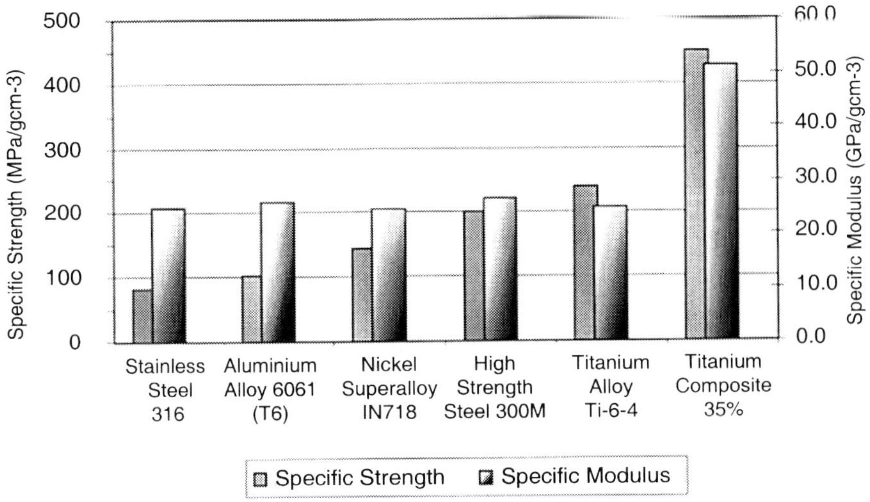

Figure 1. Graph of specific strength and stiffness vs steel, nickel and titanium.

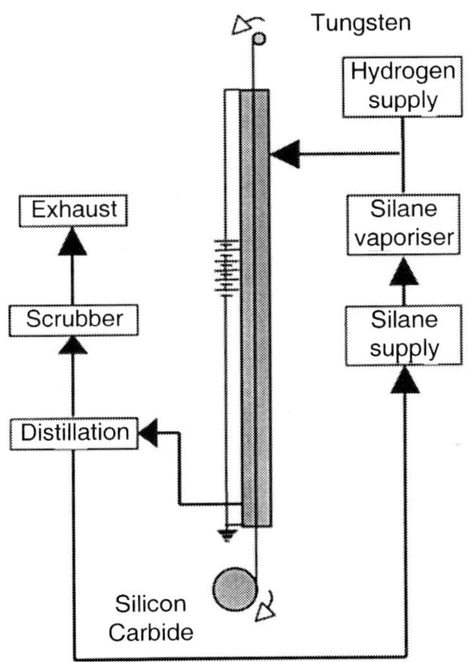

Figure 2. **Schematic diagram showing the arrangeent of the DERA Silicon carbide fibre filter.**

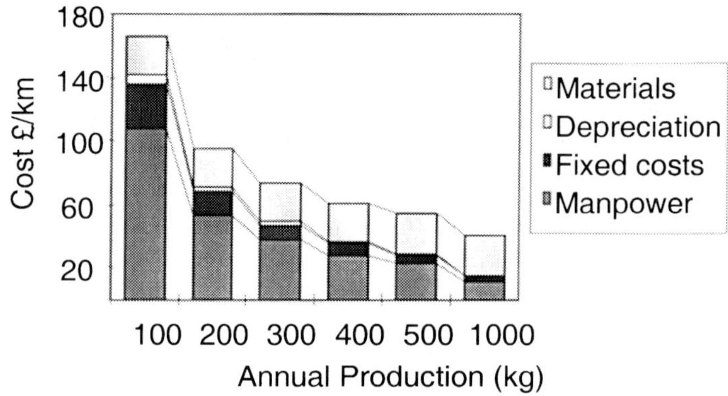

Figure 3. Cost volume curve for Sigma Monofilament.

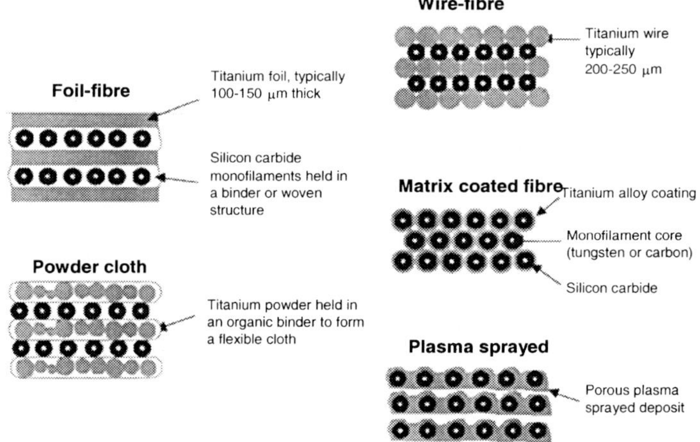

Figure 4. Schematic diagram showing the different forms of titanuim that can be used to manufacture MMC's.

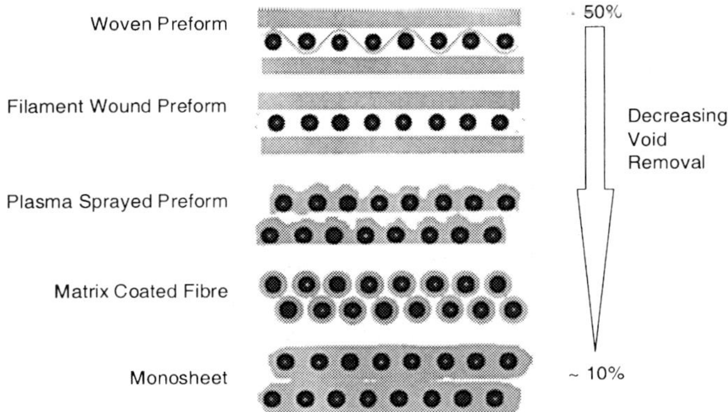

Figure 5. **Schematic diagram showing de-bulking associated with different Ti MMC processing methods.**

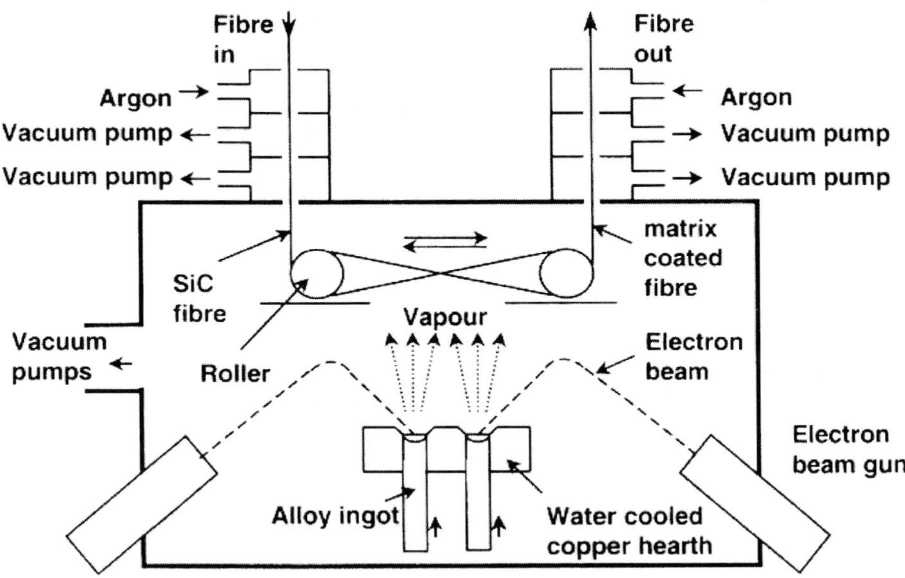

Figure 6 **Schematic diagram showing the arrangement of the DERA matrix coated fibre process.**

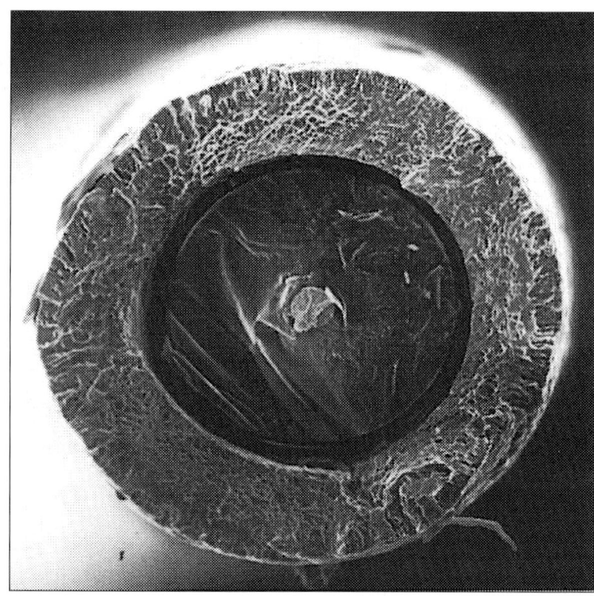

Figure 7 (a) Secondary electron image showing the general structure of DERA SM1140+ / Ti 6-4 matrix coated fibre.

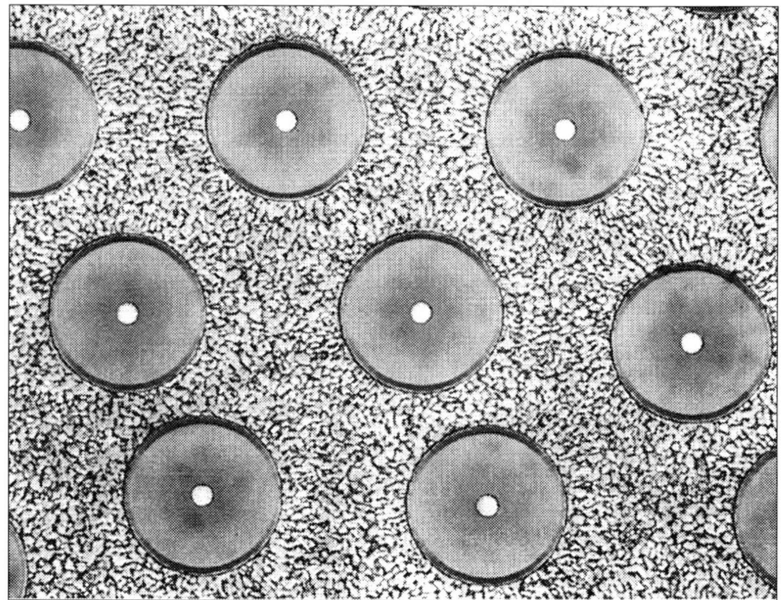

Figure 7 (b) optical micrograph showing the structure of a panel made from Ti 6-4 matrix coated fibre. In both pictures the fibre is 100 μm diameter.

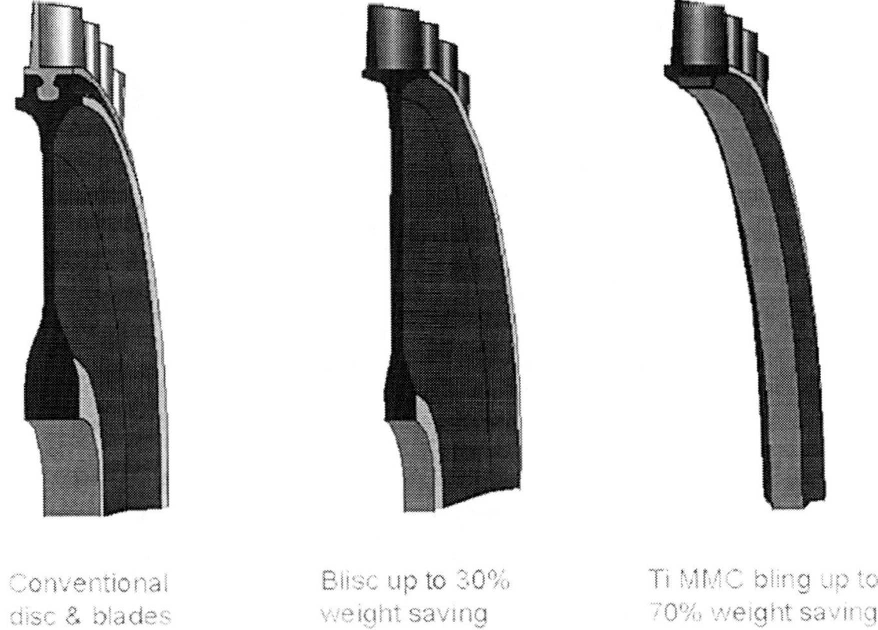

Figure 8 **Schematic diagram showing comparison between conventional disc and blade assembly, bladed disc (BLISC) and MMC bladed ring (BLING)**

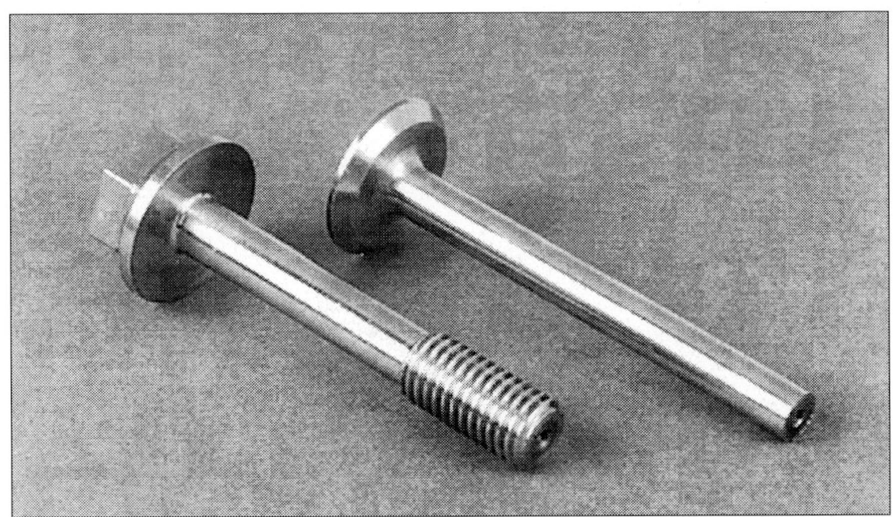

Figure 9 **Photograph showing an MMC bolt and valve stem.**

S696/008/99

Titanium technology transfer – an industrial case study

J O FOWLER
Rolls-Royce plc, Derby, UK

INTRODUCTION

Both superplastic forming (SPF) and diffusion bonding (DB) of alloy Titanium were developed by Rolls-Royce plc specifically for application in critical aero engine components, most notably the wide chord hollow fan blade for the new generation of Trent engines shown in Figure 1. The Trent is fully certified for flight and is currently specified on both Airbus A330 and Boeing 777 aircraft.

Both SPF and DB, either alone or in combination are current production processes in Rolls-Royce and have been the subject of both fundamental characterisation and an extensive programme of component development and plant investment.

SPF and DB were identified as processes applicable to the design and manufacture of improved compact heat exchangers and following a prototype development phase, a joint venture company, Rolls Laval Heat Exchangers Ltd, was formed in 1994 by Rolls-Royce plc and Alfa Laval Thermal in order to exploit the material, manufacturing and design technologies which had been developed.

SUPERPLASTIC FORMING AND DIFFUSION BONDING TECHNOLOGIES

SPF and DB are cost and weight effective methods of manufacture for both rotating parts and static structural components in aero engines, which utilise inherent structural properties of some materials to produce engineering products.

Rolls-Royce plc operates in an environment of intense competition where system design and manufacturing activity must result in cost reduction and/or an increase in the number of products or services sold.

SUPERPLASTIC FORMING

Superplasticity is a deformation phenomenon which allows some materials such as alloy Titanium and some Stainless steels and Aluminium alloys to strain to high elongations without any associated tensile instability or necking.

In order for materials to deform by what is effectively viscous flow, it must possess a fine, two phase microstructure which remains stable at the temperatures necessary to achieve high elongations at characteristically low stress levels.

DIFFUSION BONDING

During the DB process the joint faces of a part are held in intimate contact by the application of high pressures, at a temperature below the melting point of any of the sub components.

Joint faces much be processed in order to produce the required level of surface finish, cleanliness and oxide condition to allow bonding to take place, and this clean joint environment must be maintained throughout the bonding operation.

Pressure is applied by a hot isostatic method in a pressure vessel which ensures uniformity of the bonding stress over the joint.

Microscopic deformation of the surface asperities initiates bonding, and the high temperature provides the energy necessary to bring about surface void closure.

The benefits include:

- Reduced manufacturing cost
- Reduced part weight
- Short manufacturing lead times
- Reduced skilled labour content
- Improved material utilisation
- Elimination of mechanical fasteners
- Stiff structures
- Design flexibility
- Complex shapes
- High automation potential

JUSTIFICATION FOR TECHNOLOGY TRANSFER

Traditional plat/fin heat exchangers have been available for many years in wide variety of materials, but the manufacturing technologies used in their construction impose significant constraint on operational pressures. Furthermore the tooling required to generate the fin sections and the difficulties associated with joining hollow structures results in high manufacturing costs.

At about the same time that the SPF/DB hollow fan blade was being launched into production manufacture, a drive was beginning in the offshore industry to attack the weight of oil and gas production and processing installations.

Diffusion bonding and superplastic forming techniques were identified as processes applicable to the design and manufacture of improved compact heat exchangers, in order to take advantage of the potential cost and logistic gains associated with the reduction of both weight and volume of topside plant.

The technologies offer an all similar metal construction with significant advantages over shell and tube designs in terms of weight, volume and thermal performance. The space and weight savings in a North Sea gas cooling application operating at 64 barg are shown in Table 1, and illustrates both the basic weight benefit of the compact design, together with the effect of low hold up volume in operation.

Table 1. Benefits of compact design

	Rolls Laval compact design	Shell and Tube design
Material	Titanium	Titanium
Length, metres	1.1	10.0
Width, metres	1.0	1.3
Weight, empty, tons	3.7	18.0
Weight, operating, tons	4.0	28.0

The plant size reduction effect is shown schematically in Figure 2.

MANUFACTURING METHOD

Superplasticity and diffusion bonding are well known metallurgical phenomena, but have only recently been employed in relatively high volume manufacturing operations.

The process is capable of producing an almost infinitely large number of fin and header geometries and volume fractions of hollowness, which can be built up in modular form in order to achieve the required thermal/hydraulic performance.

The Rolls Laval heat exchangers currently in production utilise near net shape sub elements made of titanium alloy, the surface of which is conditioned in such a way as to promote diffusion bonding. Selected surfaces are then printed with bond inhibitor, which fixes the ultimate internal geometry of the plate/fin element. The primary pressure bulkhead bond between the primary and secondary heat exchanger surfaces is then carried out by a hot isostatic method. This results in a verifiable high integrity joint of parent metal properties, which allows the construction of a weight efficient high performance element, with the ability to take advantage of the properties of the titanium alloy. Since no dissimilar metals are used in construction, any possibility of chemical or galvanic interaction with the process media is eliminated.

The primary diffusion bonded element is then superplastically formed in a die set as shown in Figure 3 in order to produce the flat plate, uniformly strained fin structure. The crevice free channels created during SPF are available with heights varying between 2mm and 5mm. Three levels of fin waviness are also produced for each channel height. This allows optimal thermal/hydraulic design.

Elements which contain all the required distributor, collector and header inlet/outlet features are then assembled and bonded into modules, the height of which can be varied to produce the necessary heat transfer performance as shown in Figure 4 (in internally headered form).

Single modules or joined multiple modules are then assembled with exposed inlet/outlet ports, and the header welded in position. Secondary bonding and headering technologies were developed during the heat exchanger development programme, an external header design is shown in Figure 5.

THE TECHNOLOGY TRANSFER PROCESS

The aim of the technology transfer process is to offer a product to the market which can be manufactured at minimum cost, using the minimum level of working capital. In order to achieve this, a rapid and smooth transition from design concept to proven product is critical to project success.

All designs must meet cost, performance and reliability objectives. Since 70%-90% of part cost is defined at the design stage, it is essential that a close relationship is in place between engineering and manufacturing activities, in order to ensure that the design specification is compatible with process capability.

Pointers to successful technology transfer include:

- Carry out market research at the pre concept stage and continuously throughout the project
- Include customers and potential customers at an early stage
- Form dedicated interdisciplinary teams
- Call on external expertise
- Utilise simultaneous engineering techniques
- Design for ease of manufacture, installation, operation, maintenance, and disposal
- Understand the new operational conditions and environment
- Understand the commercial drivers in the new application
- Determine the critical factors and key attributes of the technology or product using a requirements capture process
- Ensure that customer critical features are identified at an early stage, and revisit throughout the project
- Employ staged product development techniques to carry out review and audit, maintain focus on the objective, and to facilitate rapid change, and visibility for all stakeholders
- Take advantage of increasing knowledge of processes to reduce variability of a product in production

CONCLUSION

An aerospace technology has been transferred into an industrial application. The SPF/DB compact plate/fin heat exchanger design enables such devices to be used for process duties that hitherto have been the preserve of traditional shell and tube exchangers, with consequent savings in both weight and space.

Rolls-Royce plc
R-R is a global company supplying current, and developing future requirements in aerospace, defence and energy markets, with facilities in 14 countries.

A broad range of products means that there are currently 53000 aero engines in service, worldwide.

In addition more than 30 navies use R-R propulsion systems, and R-R gas turbines are widely used in both the oil and gas industries and power generation.

R-R is the pioneer of gas turbine technology for aero, power generation and marine propulsion, and is today involved in major programmes for the future, including aero and industrial Trent, the EJ200 combat engine and the WR21 marine engine.

John Fowler PhD, CEng, MIM
Currently, Consultant in Materials Processing in R-R Marine Power, Derby.

Formerly, Manufacturing and Technology Manager at Rolls-Laval Heat Exchangers, Wolverhampton, and Manufacturing Technology Manager responsible for the manufacturing development of forming and fabrication processes including Superplastic Forming and Diffusion Bonding in Barnoldswick, Lancashire.

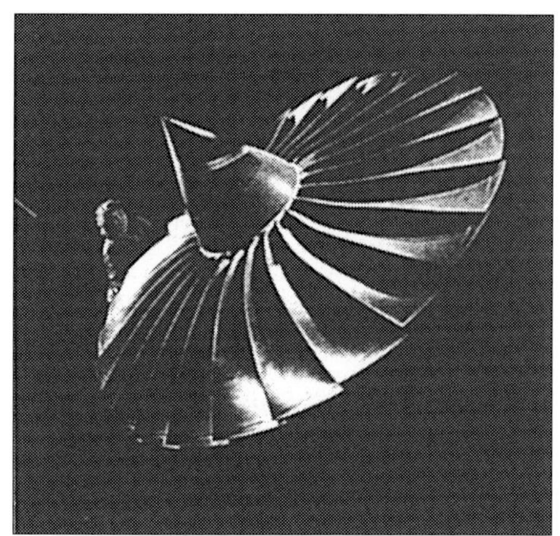

Figure 1. Rolls-Royce Trent Fan

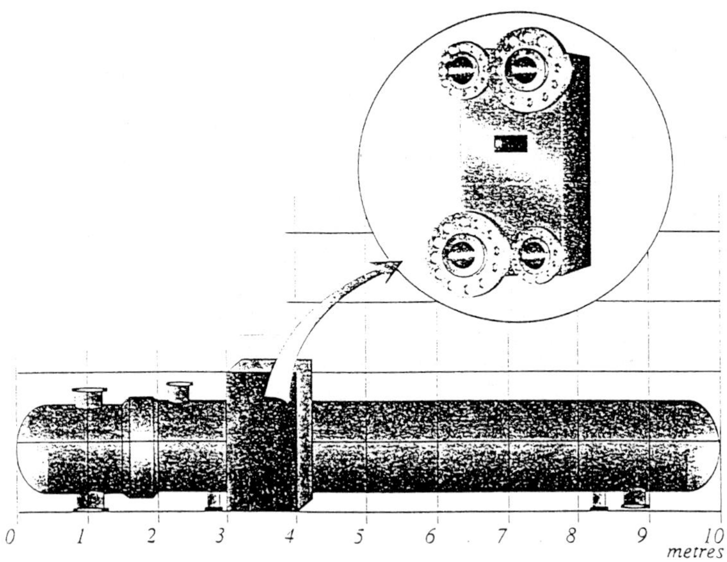

Figure 2. Plant Size Reduction

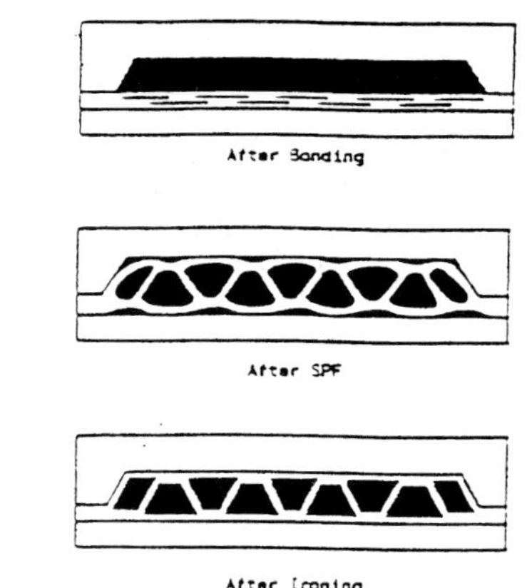

Figure 3. SPF Process

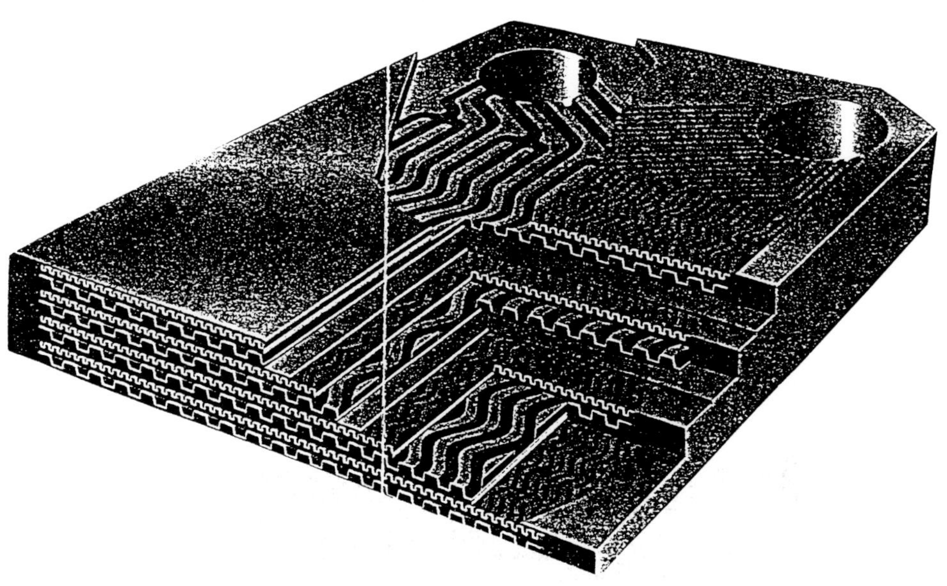

Figure 4. Segment of internally headered module design

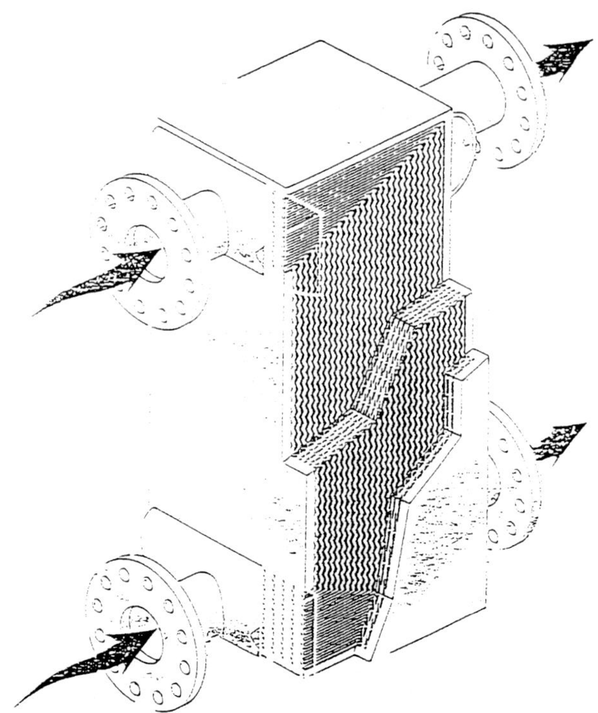

Figure 5. Externally headered single module design

S696/009/99

Titanium metal – approaching 50 and still going

I S HODGES
Timet Europe, Birmingham, UK

INTRODUCTION

Titanium celebrates it's 50th production year birthday in 2000. A metal with a remarkable and distinctive combination of physical and metallurgical properties. I just wish we'd called it something simpler rather than a name which conjures up science fiction images and the associated costs.

Titanium has had a chequered past and one of the most difficult barriers to overcome is the bad press surrounding availability and short term pricing.

YOUNG INDUSTRY WITH ROOM TO GROW

Whilst Titanium has reached the record heights of 62,000 tonnes of demand in 1998 it's still a relatively young industry and only represents about 45mins when compared to the annual production of steel. See Figure 1.

I'd like to try and overcome some of the concerns you may have by demonstrating the approach the industry is taking towards supporting and developing new applications whilst recognising our commitment to our founding industry – Aerospace.

HISTORY AND FORECAST

Figure 2 is not a view of the Rocky Mountains from our HQ building in Denver it's the history and forecast of consumption since the inception of the industry in 1950. It illustrates a couple of key factors in the evolution of Titanium. Firstly, the fairly rapid decline of Military Aerospace following the ending of the cold war in the early 90's. A market which effectively launched Titanium into production volumes and remained the mainstay from the 50's through the 80's. Secondly, the meteoric rise in usage for the sporting goods market peaking at a demand level of 4,000 tonnes which outstripped military demand in '97, something we could never have begun to imagine during the 80's. This notoriety clearly raised the profile of Titanium to new markets.

The chart clearly illustrates the boom or bust nature of the cycles, inextricably linked to Aerospace. Lead times running at 10 weeks in the trough moving rapidly to 52 weeks plus at the peak. Aerospace, begrudgingly has put up with this, primarily as the chief instigator of the demand surge. As an industry we cannot present ourselves as a credible source for long term applications in new markets unless we work with our traditional consumers to smooth out the peaks and troughs.

What the chart doesn't fully convey is the significant advances made by our customers in developing those new markets.

To demonstrate the change we've divided the history into 15 year segments, as shown in Figure 3. Industrial demand reached 10% in the 60's. During 1968 to 1983 the industrial sector growth of 10% outpaced the Aerospace growth by 3%. The last 15 years has seen a dramatic rise with Industrial at 10% vs. aerospace at 3%. These facts clearly demonstrate the growth in the industrial sector.

Why is this happening? The production process was developed around the Kroll process, VAR melting was introduced and Ti 6AL 4V with limited re-cycling became the workhorse alloy with very little development outside of Aerospace. In the late 60's and 70's, Titanium's corrosion resistance benefits became increasingly important. This lead to "new markets" Power Station condenser tubing, pulp and paper production.

By the 1980's Electron beam melting had arrived. This versatile process revolutionised the industry's ability to re-cycle scrap efficiently. More new markets emerged. Flue Gas Desulphurisation, PTA Plants, Architectural, Automotive off road, Oil, Gas and Geothermal. The Japanese were particularly active in developing Industrial markets at this time. The 90's saw further developments in Armour and Sporting goods.

1980s	**1990s**
Process Development	**Process Development**
EB Melting – Leap in scrap incorporation – Cast shapes	Expanded EB technology Industry consolidation
Sponge Vacuum Distillation – High Purity/low cost	
Investment casting – Hot isostatic Pressing	
Market Development	**Market Development**
PTA plants Flue Gas Desulfurisation Architectural uses Volume consumer/sports applications Automotive aftermarket Offshore oil and gas	Golf and other consumer goods Armor applications

Titanium is a natural for many applications, "The Titanium triangle" in Figure 4 demonstrates the selection criteria for a number of key applications.

Aerospace, positioned between strength and weight. Figure 5 demonstrates the significant move towards Titanium in wide body aircraft. Boeing's latest airliner the 777 uses around 10% of fly weight, even the latest versions of the narrow body 737 uses four times the amount of Titanium when compared to earlier versions developed in the 70's. The graph within the chart illustrates the amount of Titanium as a percentage of the aircraft backlog, the significant increase driven by the utilisation in wide body coupled with the trend towards more seating capacity per aircraft on longer routes.

TITANIUM AUTO APPLICATIONS

Automotive, on the other hand, requires facets from all three aspects. See Figure 6. Engine components requiring strength and weight, valves, springs, retainers, gudgeon pins & con rods, already well established in the off road market. Exhaust systems requiring weight and corrosion resistance, down pipes as well as silencers. Body components requiring a combination of all three, coil suspension springs, wheel nuts, brake calliper pistons and drive shafts.

Whilst the industry may see automotive as the "holy grail", I believe we're realistic enough to recognise the constraints which face us with the leap from off road tonnage's to the thousands of tonnes to support production vehicles. I'll demonstrate the industry's capacity later on but we accept a number of key factors need recognition before we can contemplate success from on increased volumes and profitability:

1) Cost reductions between 25% and 50% from today's levels with an acknowledgement of continued cost downs.
2) Partnerships with existing volume producers from other metals or green field investments with the OEM's.
3) Research and Development partnerships with external funding to develop low cost manufacturing techniques and substitution of raw materials.

Timet has entered into a number of long term (10 year) agreements with its traditional market the Aerospace OEM's. This is a step in the right direction towards satisfying the demands of Automotive.

ARMOUR APPLICATIONS

Armour requiring strength, weight and some degree of corrosion/erosion resistance is recognising benefits needed to meet the change in likely battle conditions following the end of the cold war. Equipment which needs to function in desert conditions must have reduced weight. Rapid deployment means fast transportation of men and machines. The Ultra lightweight 155mm Howitzer is shown in Figure 7. With the weight saving of Titanium, artillery gets the firepower of a 155mm for the weight of a 105mm, rapidly deployed with heavy lift helicopters. Further applications will see extensive use of Titanium.

ARCHITECTURAL APPLICATIONS

Architectural applications initially attracted by the aesthetic aspect of the metal brought weight benefits to designers looking to reduce the mass of the supporting structure as well as the opportunity to try out new materials. The Guggenheim museum in Bilbao used 60 tonnes of thin gauge sheet in a unique panel configuration which complimented the location of the building. Figure 8(a) shows the Van Gogh museum in Rotterdam and Figure 8(b) the National Scottish Science Centre in Glasgow currently under construction. Both these examples are similar to the Guggenheim, using the light refracting qualities of the metal as well as the easy fabricability of thin sheet. Projects currently in the pipeline are being designed using anything up to 100 mt of Titanium. The very latest concepts are planning on using three times this amount.

OFFSHORE APPLICATIONS

Oil and Gas

With one major riser project successfully installed in the North Sea interest is increasing (Figure 9). By alloy selection to enhance Titanium's resistance to SCC at low sea temperatures Titanium becomes a natural candidate for production as well as drilling risers. Once again these projects are utilising hundreds of tonnes. Whilst topside applications like Service and Fire water systems have been around for a few years, the innovative high pressure heat exchangers which takes up a fraction of the space and weight of traditional shell and tube exchangers is still relatively new. Both the reduced weight above and below the water allow designers to reduce the platform size with the associated cost benefits. See Figure 10.

I've mentioned a handful of the many new applications, some in service others conceptual. Whilst we see these applications as our true growth market we must also recognise the challenge facing producers. Any failure to support programmes at such an early stage would see either traditional materials return or projects stopped in their tracks. In the past our credibility was severely dented by our inability to get enough ingots into the production pipeline. The two elements behind ingot production were sponge production and double or triple VAR melting. Over the last few years following a long period of loss and under investment the industry has invested a third of a billion dollars bringing on new capacity aimed at reducing the dependence of both factors. The vast majority of new melting technology has been aimed at scrap re-cycling. My own company has installed an additional 10,000 tonne furnace, adding to the existing 7,000 tonnes. Projects currently underway will add another 10,000 tonnes by 2001. See Table 1.

With only 40% utilisation it clearly gives us a short term headache trying to fill our factories, but a medium to long term message called adequate capacity to support. The situation for sponge production is slightly different. The last "new" sponge plant was finished in 1993 with a capacity of 10,000 tonnes of Vacuum distilled sponge. Sponge is not regarded as a restrictor as many lines are now mothballed. This adds up to an ability to satisfy ingot production for the foreseeable future. The challenge will be downstream processing, however nearly all Titanium production can be achieved using existing steel production equipment, and provided we think globally an infinite capability exists.

POTENTIAL GROWTH

Finally, the future. If we sit back and let the Airline growth figure of just under 5% drive us we should reach 100,000 tonnes by 2013 (Figure 11). This equates to evolution. Based on historical trends I have no reason to doubt that figure. This will result in years of under utilisation and wasted potential. We need to exploit new opportunities and speed up that pace of change. Perhaps a growth rate of 10% is not out the question resulting in 250,000 tonnes. This equates to revolution.

	Capacity 000's of MT		Utilization %	
Sponge				
North America	21.0		60%	
Japan	26.0		75%	
FSU	69.0		40%	
China	3.0		85%	
Europe	-		-	
	119.0		52%	
	VAR	EB	VAR	EB
Melting				
North America	59.0	34.0	70%	40%
Japan	21.0	-	75%	-
FSU	52.0	2.0	25%	25%
Europe	11.0	-	70%	-
China	4.0	-	25%	-
	147.0	36.0	54%	40%

Table 1. World capacity and utilization.

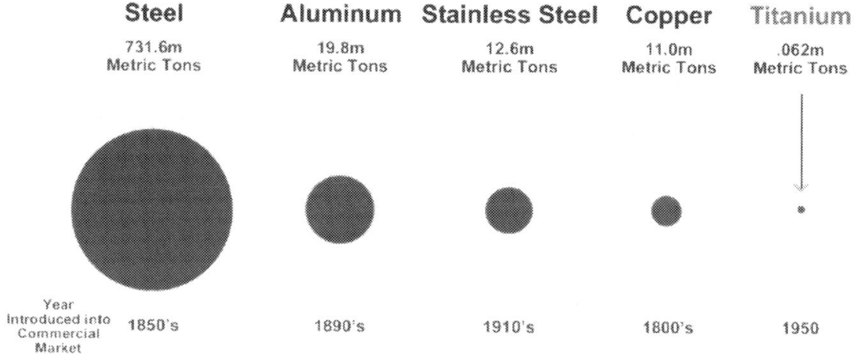

Figure 1. Titanium industry growth.

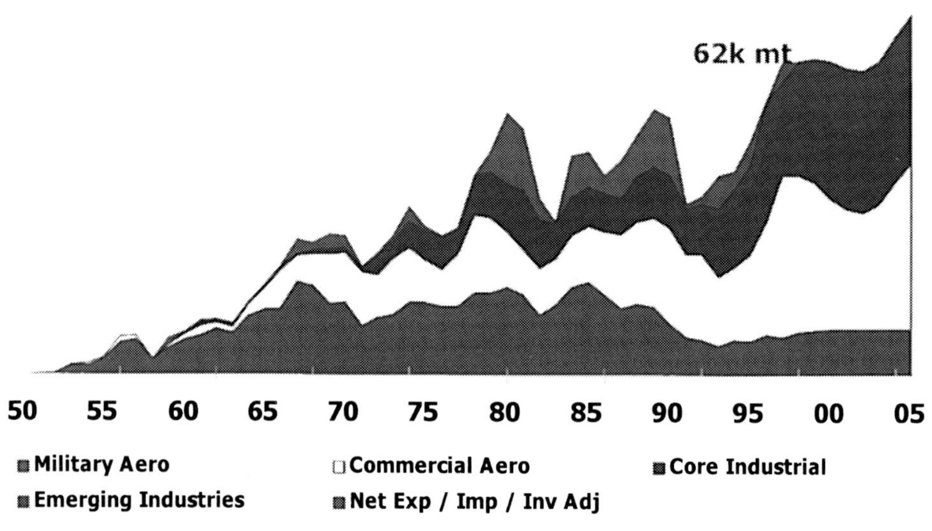

Figure 2. History and Forecast.

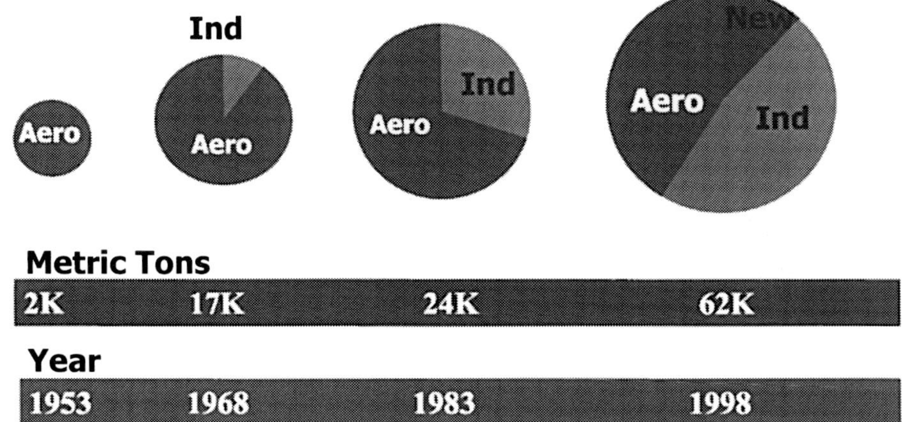

Figure 3. World Titanium market growth.

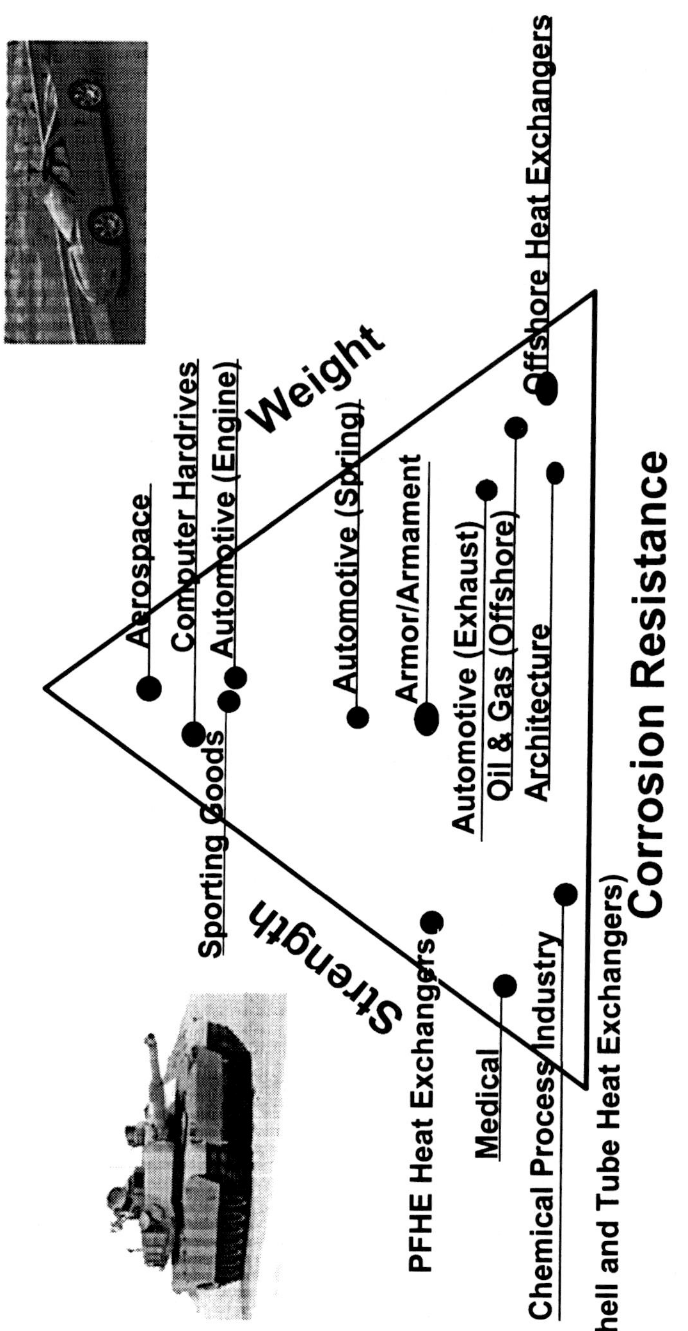

Figure 4. 'The Titanium triangle'

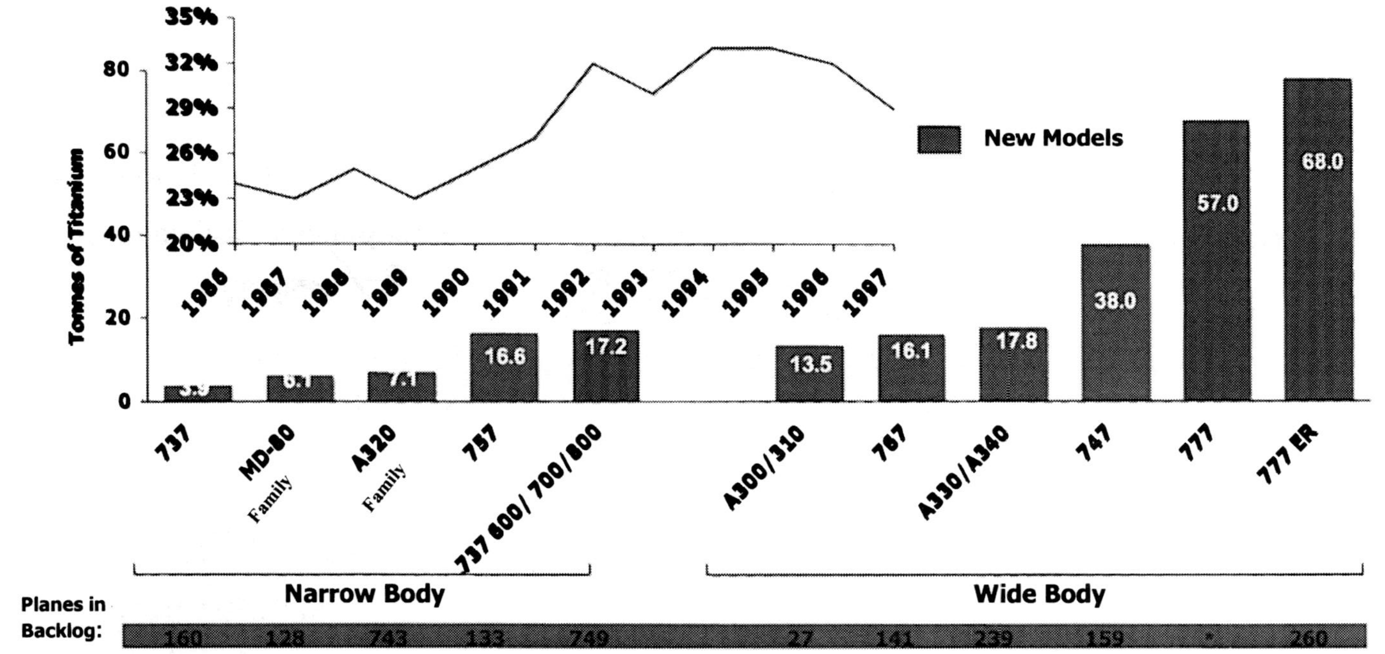

Sources: ***The Airline Monitor*** **and Company estimates as of 12/31/97.**

*** Backlog for the Boeing 777 ER included in base model.**

Figure 5. Titanium usage per plane (fly wt)

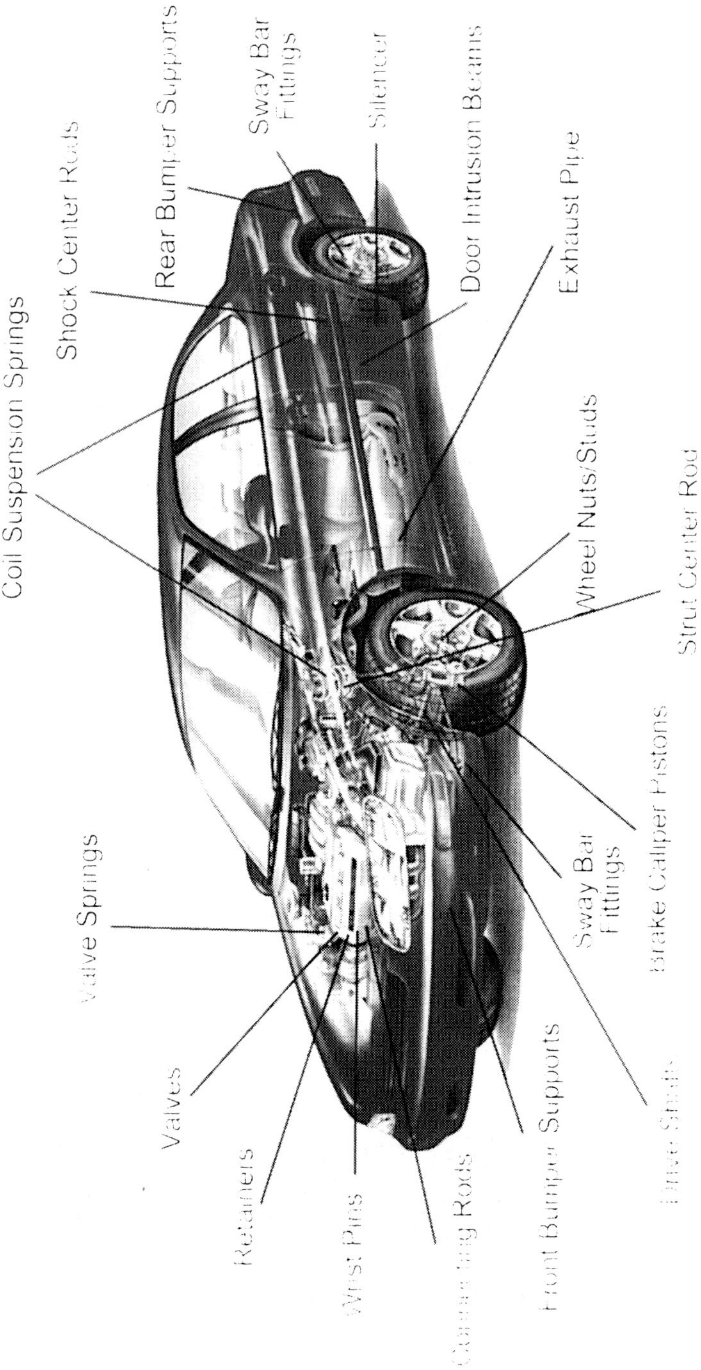

Figure 6. Titanium auto applications.

Figure 7(a). Armour applications.

Figure 7(b). Armour applications.

Figure 8(a). Van Gogh museum in Rotterdam.

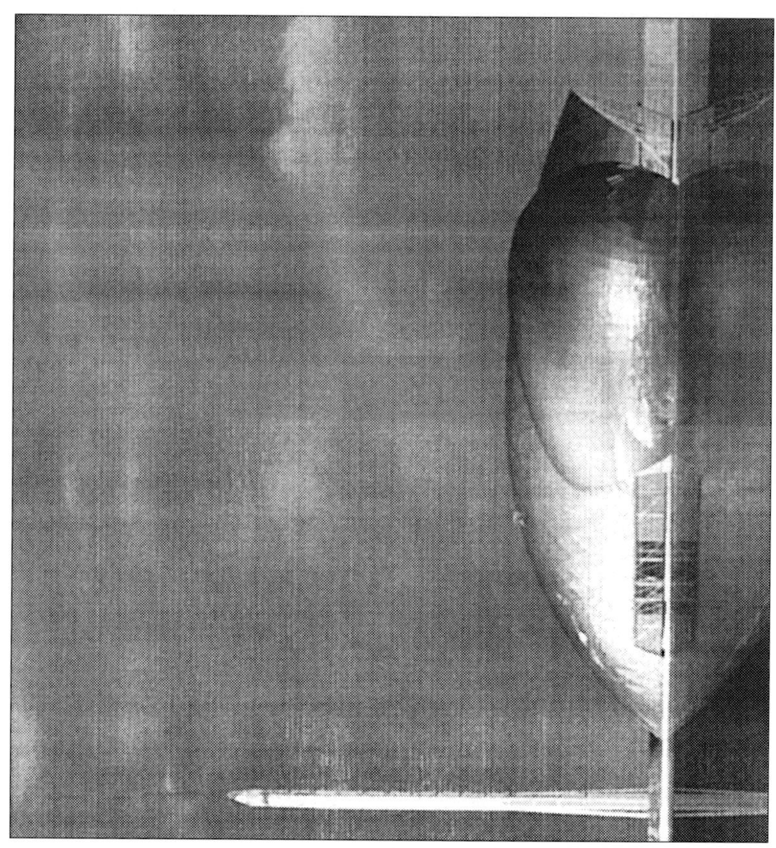

Figure 8(b). National Scottish Science Centre – Glasgow.

Figure 9. One deep hole rig with Titanium drilling and production risers, along with associated topside equipment, could use up to 750 mt of finished produce, yielded at 50% = 1,500 mt of ingot.

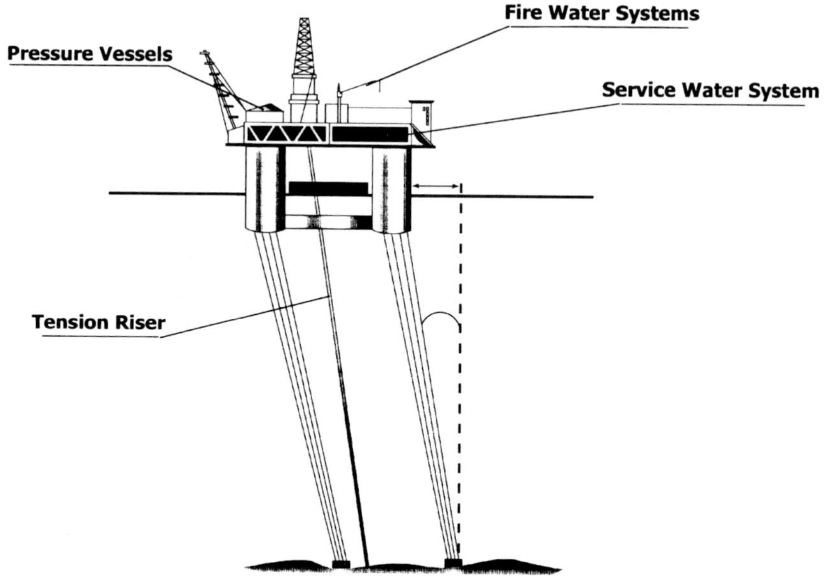

Figure 10. Tension leg platform with stress riser.

MT in 000's

250
200
150
100
50
0
2
17
24
62
100
250
1953
1968
1983
1998
2013
Actual
Evolution = 3.5 %
Revolution = 10 %

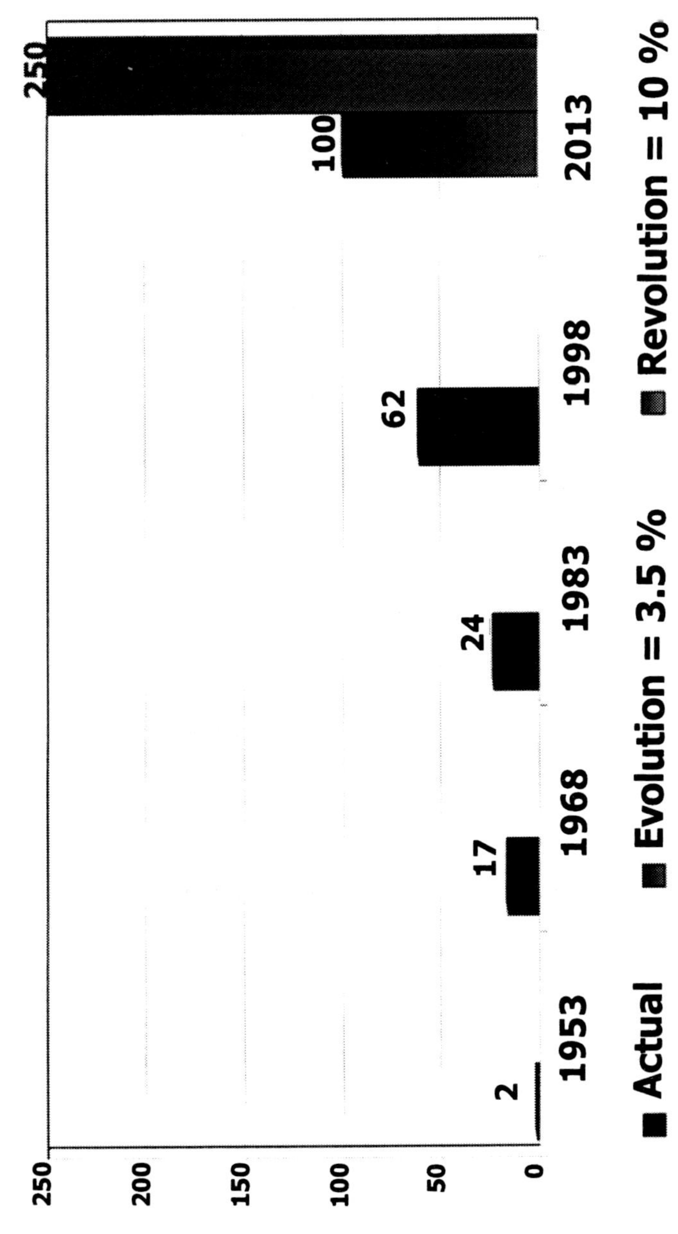

S696/010/99

Titanium in Formula One racing car construction

A J SMITH
B3 Technologies Limited, Guildford, UK

INTRODUCTION

The use of titanium and its alloys in the construction of Formula One racing cars has had a somewhat chequered history. Over the last few years however, its use has been extended into new and more challenging areas, with increasing use of novel design and manufacturing techniques.

Although undoubtedly an expensive sport, Formula One manufacturers are under ever increasing pressure to produce leading edge designs which are both efficient and cost effective to manufacture, in terms of both direct cost and time.

This paper will discuss various race car components manufactured from titanium, and demonstrate how the differing manufacturing routes can be used to produce efficient, cost effective end products. By necessity, it only covers work that I have been involved with, but is indicative of the "state of play" throughout the industry.

The aim of the Formula One car designer is to produce a vehicle whose components allow the car to perform to the maximum. To this end, components need to be both light and stiff. This maximises the power to weight ratio of the car, and allows the movement of items such as suspension to be controlled, whilst minimising flexing of the components themselves. Thus, over the years, we have seen more and more of the cars being manufactured from materials such as carbon fibre composites, which exhibit exceptional specific strength and stiffness properties. However, whilst undoubtedly suited to components such as chassis, suspension members, bodywork and wings, their use is limited to those components which can make use of the inherent anisotropic properties of such materials. Steels, particularly the ultra high strength low alloy steels such as SAE 4130, 300M and the maraging steels have been used extensively for components such as suspension uprights, springs, fasteners and brackets. However, with a density of approximately 8 g/cc, the weight penalty for using steel can be considerable. Aluminium alloys possess a much lower density, but their usefulness is limited to low strength/stiffness applications.

Within the extremes of composites and steels, lie the titanium alloys. With considerably lower densities than steels, they also have only half the stiffness. For critical components therefore, the designs must be optimised to "regain" the lost stiffness without adding too much material, and hence weight. A notable exception to this, is the use of titanium in springs and "flexures", which will be discussed later.

TITANIUM FABRICATIONS

Suspension Components

As was mentioned in the introduction, titanium was the cause of great concern during the early 1980's. It was during this time that titanium was first used in the manufacture of suspension components and control pedals.

When titanium alloys are heated, they have a great affinity for oxygen, nitrogen and hydrogen. These atomic impurities embrittle the base metal, leading to poor fatigue performance. The problem is particularly relevant during welding, and requires great care to ensure that the material is protected from these impurities. Conventionally, this has required the use of inert gas shields around the weld site. However, this can be very difficult to achieve, and requires great care on the part of the welder. It was this embrittlement of welds that caused a series of suspension failures, and probably the failure of a brake pedal during the early 80's. For almost a decade, titanium was avoided by formula one designers for components that required welding.

In the late 1980's, GTO, the English design and manufacturing arm of the Ferrari Formula One team produced a fabricated titanium roll hoop using Ti-6Al-4V alloy. Although successful, it was very time consuming to manufacture, due mainly to the difficulty in forming sheet material. TIG welding of the constituent panels to form the completed structure was carried out using just the shielding gas of the welding torch.

The manufacture of highly stressed titanium fabrications such as uprights, required a considerable improvement in the techniques employed for the roll hoop mentioned above, in terms of performance and manufacturing efficiency. The "upright" on a racing car is the component that holds the axle assembly, and apart from mounting the brake calipers, provides the attachments for the wishbones and pushrods thereby transmitting the wheel loads into the chassis. A modern Formula One car produces cornering loads of up to 5g, with similar loads in braking and acceleration. This compares to the 1g of braking load produced by the best road car. Conventionally made from SAE 4130 steel, we introduced the first successful titanium upright to Formula One. Again, manufactured from Ti-6Al-4V, the upright was manufactured using a variety of techniques. To reduce the lead-time for complex items such as uprights, we utilise techniques such as wire eroding to produce the air passages around the central spool required for brake cooling. In the past, these passages were created using "vanes" individually welded in place. Over the last few years, more and more teams have invested in wire eroding technology. The application of this technique is being pushed to the limit; not only to produce components, which cannot be manufactured by any other means, but also to reduce the lead-time of more traditionally manufactured parts.

Suspension mounting points are attached to the central spool using fabricated skins. Once developed from the component drawings, the patterns can be digitised and the parts laser cut from titanium sheet, with the addition of laser etched markings for identification of fold lines etc. This production route can save in the order of two man-days over cutting by hand the numerous templates required for an upright. In addition, the laser etched fold lines do not result in the stress raisers caused by scribed lines. The suspension mountings themselves are manufactured from machined titanium bar, and are attached to the central spool via the skins, which must be formed to the required shape. This process in itself is not easy, as the inherent "spring back" of titanium sheet when formed, and its inability to form tight radii mean that the forming must be carried out hot. This again poses the risk of contamination, and steps must be taken to minimise the effect.

The welding of the upright requires particular care. Any given upright design involves a considerable amount of welding, all of which must be carried out to the highest possible standards. It is not sufficient to rely on the appearance of the welds in terms of estimating full penetration and the lack of weld pool contamination. Over the years, B3 Technologies has gained considerable experience in manufacturing safe, efficient titanium fabrications, with all our fabricators CAA approved. This is combined with our own design of inert gas welding chambers and techniques for ensuring the quality of the resulting welds, including surface preparation and NDT testing.

Once machined, titanium uprights provide considerable weight saving over an equivalent steel component. This is particularly important for "unsprung" suspension components. However, to maximise the advantage, the upright must be carefully designed such that the resulting system stiffness is not sacrificed in the search for reduced weight. Titanium uprights have now been run by various teams, with no significant problems. Indeed, they seem to suffer less from in service fatigue cracking than their steel equivalents. A typical titanium upright is shown in Figure 1.

Gearbox

Conventionally, gearboxes in motorsport are manufactured from cast magnesium alloys. This material and the production route result in a number of problems. The gearbox of a Formula One car is a major load carrying structure which provides all the attachment points for the rear suspension, and feeds the resulting loads into the chassis via the engine, and as such, should be as stiff as possible. With this in mind, Ferrari Design and Development (the forerunner of B3) developed a steel fabricated gearbox casing, as a pre-cursor to the introduction of a titanium version. The resultant design was considerably stiffer than a magnesium box, with little weight penalty, and with the introduction of the titanium version, the advantages were further enhanced. Manufactured from machined titanium plates and formed sheet, the gearbox proved very resilient, apart for one notable exception. During the production run, welding of some of the cases was carried out by an outside source. Not used to the requirements of welding titanium, the case suffered from major cracking along weld lines. Outwardly, the welding appeared fine, but sections revealed a lack of penetration, and contamination of the welds. This stressed how important it was to maintain absolute quality during manufacture.

An additional advantage of the fabricated gearbox method, was the ability to alter such things as suspension pick-up points. During the racing season, it is quite common to optimise the suspension geometry, and a fabrication allows fundamental changes to be incorporated relatively easily. For a cast magnesium component, such changes are difficult and costly, requiring offset brackets at best, and redesign of the casting patterns at worst.

Joining Techniques

Apart from manual TIG welding, numerous other techniques are used for joining titanium components to themselves, or other materials. A widely used method is that of electron beam (EB) welding. Titanium is readily welded by this technique, which benefits from the process requirement that all welding must be carried out under high vacuum, hence minimising the risk of contamination. Weld efficiencies equal to that of the parent material are possible, and the technique allows considerable scope for weight saving and ease of manufacture. Complex assemblies such as anti roll bars are possible, using multiple EB welds. These avoid costly and heavy mechanical forms of assembly such as splined joints. Component production may be simplified and/or significant weight savings can be realised by manufacturing components in multiple parts, then EB welding them to form the finished component. Although the Ti-6Al-4V alloy is that most commonly used in fabrications, other more exotic alloys such as Timet 10-2-3 have also been used with success.

It is often be desirable to join titanium with different materials such as other metals, composite structures or ceramics. In these cases, methods other than welding must be employed. Techniques used include adhesive bonding, vacuum brazing, and diffusion bonding.

Adhesive Bonding

In 1991, Benetton Formula were one of the first teams to produce a steering column produced from titanium end fittings adhesively bonded to a carbon composite tube. The resulting assembly achieved significant weight savings over a conventional steel design, and provided increased torsional stiffness. The column was also quick to manufacture. Obviously, with such a safety critical item, great care was taken to ensure the integrity of the bonded joint. Although the joints were not highly stressed, the bonding operation had to be carried out with meticulous attention to detail. Surface preparation and cleanliness and bonding methodology are critical if the joint is to perform as intended.

Whilst at Ferrari, John Barnard (Owner of B3 Technologies) designed a new kind of suspension joint which removed the need for spherical joints. Known as "flexures", these were essentially hinges that allowed the vertical movement of the wheel. Conventional spherical joints used in items such as wishbones suffer variability in friction, particularly in areas such as the rear of the car, where joints can get very hot. This leads to variability in performance of the suspension system. Since at that time, suspension was still manufactured from steel, the original flexures were made from the same material, allowing fabrication by the normal TIG welding route. However, the operation of the flexure requires a low bending stiffness, to reduce the load required to move the wishbone, but a section size sufficient to resist buckling. During heavy braking and acceleration, suspension legs experience compressive loads in the order of 20 –30 kN. The obvious alternative was to manufacture the flexures from titanium. With half the stiffness of steel, it also has the same strength as the SAE 4130 steel used in conventional suspension. Since the suspension legs were still of steel, the only efficient joining technique was adhesive bonding. An early prototype flexure element is shown in Figure 2. With the much higher stress levels experienced in suspension over the steering column mentioned above, the bonding operation became even more critical.

However, with the correct surface treatments, and attention to detail, the bonding route can be successful.

In the continual pursuit for reduced weight and increased stiffness, many current Formula One cars use composite suspension components. The use of carbon fibre composites allows the design and production of components such as wishbones that are simply not possible with metals. However, the points at which the suspension members attach to the chassis or the uprights, invariably must be made from metal, (usually titanium) whether they are flexures, or conventional joint housings. Adhesive bonding has thus become a very important production route for highly loaded, safety critical components.

One problem area that exists however, is the non-destructive testing of complex bonded joints. Many techniques have been used, from ultrasonics to x-ray and laser shearography, but I have yet to find a truly useable solution for the Formula One industry. The problems lie in the complex geometry of the joints themselves, the multiple interfaces in the joint (particularly in composite structures), and the types of defect that are potentially present. Nearly all problems I have encountered in such joints stem from poor surface preparation, which yield "zero volume defects" and are very difficult to detect. Ultimately, confidence comes from careful manufacturing techniques and mechanical testing of the components throughout there life.

Vacuum Brazing and Diffusion Bonding

Vacuum brazing can be a useful technique to join titanium to differing substrates, especially when adhesive bonding is not appropriate for reasons of either joint design or operating conditions e.g. high temperatures. A requirement for a hard abrasion resistant surface to titanium led to the use of vacuum brazing to join zirconia to a machined titanium base. Although titanium will not tolerate many metallic contaminants such as iron which leads to extremely brittle interfaces, brazes based on copper and silver can be successful. The component mentioned above was originally manufactured from case hardened steel, but suffered from premature wear, and operated in a hostile environment. The composite titanium/ceramic version showed great promise, with wear reduced to negligible levels, but slight processing difficulties lead to the project being abandoned. However, with a little perseverance, I am sure this technique would be successful. Vacuum brazing has also been used to manufacture hollow titanium components with wall a thickness of approximately 2mm. This can be achieved by initially machining the internal detail of the component in titanium plates, brazing the blocks together, then final machining the external surfaces.

Diffusion bonding is a technique whereby certain metals in intimate contact, will fuse together with migration of atoms across the joint interface. The resultant "bond" has properties identical to the parent metal. Any surface contamination prevents the process occurring, but titanium has the ability to dissolve any such contaminants into the body of the material at the temperatures used for bonding. Figure 3 shows an anti-roll-bar assembly originally designed for EB welding. As part of the assembly process used to assist the sub-contract welders, the shaft and beam are shrunk fit together in order that the orientation of the beam with the spline on the shaft is maintained. Initial experiments have shown that this transition fit is sufficient for diffusion bonding to occur readily. The benefits of this is that the components can be pre-assembled, and the diffusion bonding can be carried out in a normal vacuum furnace, without the use of fixtures or the application of external pressure. One potential concern with this production route is that, at least with the configuration shown, it is not obvious that any diffusion bonding has actually occurred without sectioning the piece, or proof testing the component after processing. However, the technique does offer a potentially cost effective method of joining shafts to other features, with a high degree of mechanical integrity.

MACHINED COMPONENTS

A common problem we come across, is the reluctance for sub-contractors to machine titanium, particularly the beta rich alloys such as 10-2-3 and Beta-C. This seems to stem from a perceived idea that it is particularly difficult. We have found that as long as the approach is correct, then few problems should occur. Machine tools should be as rigid as possible, and the cutters used designed for the job. This usually means solid carbide cutters, or those coated with TiCN or one of the other proprietary coatings. Cutters will also not last as long as those used for other materials. This is most evident when tapping small threads, especially if the thread is deep relative to its diameter. We have found that the only solution here is to use taps for a very short period – sometimes only a couple of applications – and to use taps specifically designed for titanium with a good tapping fluid or gel. It also helps to have a good relationship with your local spark eroder!

Wire and spark eroding are manufacturing methods, which can have a dramatic effect on the cost of making a component. As has been mentioned, the vanes on uprights are wire cut, saving considerable time in fabrication. The technique has been used this season to manufacture highly loaded suspension components that would otherwise have been impossible to manufacture. Spark eroding is very useful for not only removing broken taps, but for features such as hexagonal sockets.

Apart from the usual brackets etc., which have been manufactured from solid titanium over the years, titanium offers some attractive properties for other more highly loaded parts. As for the flexures, the low stiffness compared with steel, combined with high strength levels, makes titanium an attractive material for use in torsion springs. This form of spring medium is becoming more popular with designers to replace the more conventional coil springs, due to their "packaging" potential in the ever diminishing volumes available in the chassis. Of the commercially available titanium alloys, Beta-C offers one of the lowest modulus combined with high strength once aged. As such it has been used very successfully for such springs. This is despite its reputation for variable mechanical properties with processing variability and from melt to melt. No such problems have been encountered, although for such critical components, it is only wise to minimise batch variation during a production run, and monitor heat treatment using test pieces that can be mechanical tested to verify results.

Other alloys used for highly stressed components include 10-2-3, which has been used to good effect in suspension rockers where its high fatigue limit is particularly useful. Figure 4 shows such a rocker manufactured a few seasons ago. The area of particular concern was the drive dogs, which engaged with the torsion spring, and the use of this particular alloy allowed for a successful design. Other components manufactured from titanium include axles, which when combined with titanium uprights, significantly reduce the unsprung mass of the front suspension system. The correct alloy selection, processing, component design and manufacture has resulted in a component, which will last at least a season of racing with no problems. This compares to the first titanium axles used in Formula One that had to be discarded after every race!

When designing and producing highly stressed components that do not have the usual high safety factors used in other industries, care must be taken to use material whose origins are known, and to be aware of the structural implications of using a particular form of material to produce the component in question. For instance, large diameter bar should not be used to produce components whose maximum stresses occur at small diameters. Despite being costly in waste material, the coarse grain structure at the centre of large bars will not give the maximum potential properties. In cases such as these, forging is often a cost effective method to reduce waste and improve component integrity by optimising grain flow and size.

SURFACE FINISHING

A problem that must be addressed when using titanium, is the potential for galling, particularly with fine threads. It may also be evident in situations where rubbing or sliding exists. The problem is exacerbated if both mating components are titanium. Fortunately, there are now a variety of surface treatments available that can ease, and in some cases remove the problem. Essentially, they involve producing a coating or surface layer that prevents intimate contact and/or reduces surface friction. Depending on application, coatings such as Titanium Nitride (TiN) and Titanium Chromium Nitride (TiCN) can be useful. Treatments that have a more fundamental effect on the surface layers of the titanium include Ion-Slip™ from AEA Technology, and Pulsed Plasma Nitriding. The former process has the advantage that the processing temperature is low, in the order of 150°C so causing minimal distortion. Plasma Nitriding is a much higher temperature process, but has the advantage that the effective layer is relatively deep. This offers the potential to use titanium in situations where high point loads are experienced, an area where titanium does not generally perform well. Too often, an otherwise sound design has failed because the chosen treatment achieves the required properties on the surface of the component, only for the titanium substrate to "collapse" beneath.

Fatigue life can be dramatically improved by the use of controlled shot peening, as for other metal components. What must be borne in mind however, is the detrimental effect iron contamination can have on the properties of titanium, particularly in fatigue – the exact area shot peening is designed to improve. The solution is either to use ceramic shot, or to de-contaminate the component after initial peening using glass bead.

CONCLUSION

Titanium is being used extensively in the construction of Formula One racing cars, where its particular balance of strength and stiffness and allows efficient, low weight designs in ever more highly stressed situations. The use of modern manufacturing techniques can not only make the designs possible, but also cost effective.

Figure 1 – Titanium Upright

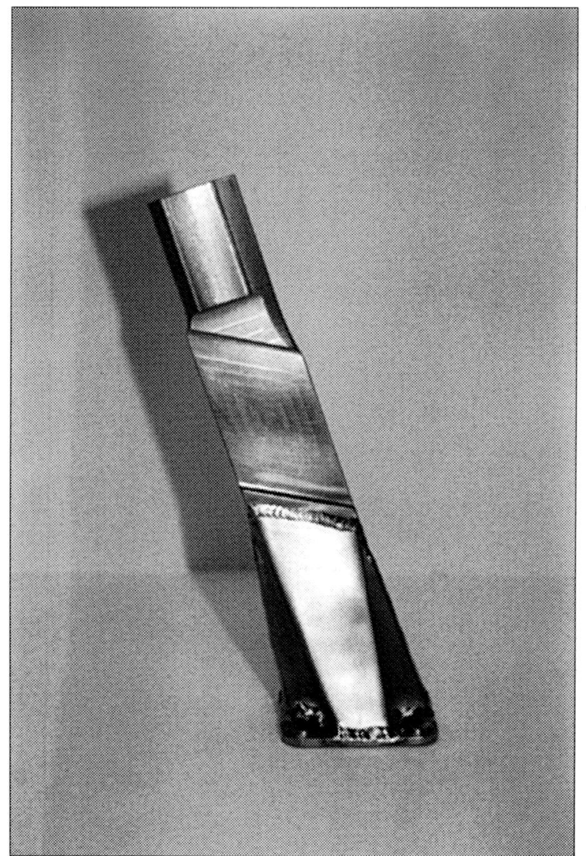

Figure 2 – Prototype Flexure Element

Figure 3 – Diffusion Bonded Anti Roll Bar Assembly

Figure 4 – Timet 10-2-3 Suspension Rocker